Biography

Susan Prescott is a practising specialist paediatrician at the leading children's hospital in Perth, a research strategy leader at the Telethon Kids Institute, and a Winthrop Professor at the University of Western Australia. She is the founding president of the Developmental Origins of Health and Disease (DOHaD) Society of Australia and New Zealand, a society dedicated to providing strong evidence, education and advocacy around the importance of a 'healthy start to life', and is also on the council of the International DOHaD Society. Professor Prescott is internationally recognised for her research in the area of allergy and early immune development, and is the founding leader of the International Inflammation Network 'in-FLAME', focused on promoting early immune health to reduce the risk of many chronic inflammatory diseases in the modern world.

Origins

Early-life solutions to the modern health crisis

Susan Prescott

UWA PUBLISHING

First published in 2015 by
UWA Publishing
Crawley, Western Australia 6009
www.uwap.uwa.edu.au

UWAP is an imprint of UWA Publishing
a division of The University of Western Australia

THE UNIVERSITY OF
WESTERN AUSTRALIA

National Library of Australia
Cataloguing-in-Publication entry

Prescott, Susan L, author.

Origins : early-life solutions to the modern health crisis

9781742586700 (paperback)

Health. Health planning. Health—Social aspects—21st century.

Quality of life.

306.461

Typeset by J & M Typesetting
Printed by Lightning Source

This book is dedicated to the future generations,
and to all those who work tirelessly,
now and ever,
from every walk of life,
for a better future.

CONTENTS

Preface

This is a book about health. But it is also about our future. It is about understanding our past, so that we can improve the health of our own future and the next generation's, in every regard. At its core, the purpose of this book is to bring a greater awareness of the critical importance of improving conditions in early life for long-term health and longevity. In other words: how well we start life will determine how well we finish it.

The environment has shaped us each, from the first moments of our personal lifetime, and over millennia through effects on the genes of our ancestors. It is by understanding our intimate relationship with the natural world, and the adverse effects of the modern world on our biology, that we can best address the modern health crisis. Our relationship with the environment begins at our conception, and even before that, with the health of our parents. Adverse conditions during critical stages of our development can have a profound effect on our body structures, functions and even our developing behaviours. In many cases, the consequences may not appear until much later in life, through effects on our biological reserve, and our capacity to deal with life's challenges. This also means that to truly overcome the rising rates of modern disease, we must target prevention from the first moments of life, addressing the many modern risk factors shared by so many chronic diseases.

Yet, we can so clearly see that human health is intimately related to the health of our environment and the health of our societies. This means that we must ultimately go beyond the biology to solve our health problems. The many global challenges across so many domains cannot be solved in isolation, and will ultimately require a more coordinated, collaborative, cross-sectoral approach. This needs a broad vision, and a long vision, with a deep understanding that the solutions of tomorrow must begin today. The health of tomorrow will depend on the choices we make together today. In this, we must have good evidence and we must understand our biology before we can go beyond, to solve the greatest challenges of our era.

Acknowledgements

That this book is written at all is a testament to the work of Professor David Barker, who inspired so many to follow his quest to understand the significance of the 'early-life origins' of health and disease. I am deeply grateful to Professor Lawrie Beilin who, in 1986, early in my research career, opened the door to this journey and gave me a firm grounding in the importance of nutrition, lifestyle and a holistic approach to health through disease prevention. It was through his early connections to the Barker group that I first became aware of the 'early-life origins of disease' hypothesis, long before I appreciated its significance.

In more recent years, I am grateful to the leadership at University of Western Australia, particularly Vice-Chancellor Paul Johnson, Deputy Vice-Chancellor Robyn Owens and Professor Alistar Robertson, for making 'Developmental Origins of Health and Disease' (DOHaD) a priority initiative, and encouraging me to establish the DOHaD Society of Australia and New Zealand as its founding president. The impetus for this came after attending the Worldwide Universities Network (WUN) first Public Health Conference in Shanghai (2011), which focused on *'Early Life Opportunities for the Prevention of Non-communicable Disease (NCD) in Developing Countries'*. There, I was inspired by the work of Professor Mark Hanson and Professor Sir Peter Gluckman, and the seeds of this book were sown. I am also very grateful for the support and encouragement of Professor John Hearn, chief executive of WUN.

I owe a great debt of gratitude to the countless researchers from so many fields, upon whose work the DOHaD hypothesis has been built into a substantive reality, and from whose many publications I was able to draw the material for this book.

I wrote this book during stolen moments on my many travels, and I wish to give thanks to all those who shared their thoughts and inspirations along the way. And a very special thanks to fellow authors Mark Haynes Daniell and Alan C. Logan for their encouragement.

As ever, this would not have been possible without Professor Terri-ann White, Director of UWA Publishing, who believed in this project even before I suggested it. I can never thank Terri-ann enough for the opportunities she has given me, and for her mentoring and friendship. I also extend my deepest thanks to all of the staff at UWA Publishing, and in particular to Suzannah Shwer for her editorial assistance in streamlining the manuscript from an original much larger body of work.

To my dearest love and husband, Craig, who provided material for many sections of this book through his own research as he studied for his Masters of Public Health: thank you, as always, for everything. And to the generations of my family, past and present, for the inspiration to work in health, always with a higher vision of the possibility for societal transformation – thank you for nurturing my purpose from the first moments of my life.

1

A pilgrimage

It is February 2001, and I am on my way to the first international conference on the 'early-life origins' of health and disease, in Mumbai. I'm soaking up my first impressions of this wonderful, crazy place on the journey from the airport. No fear; no seat belts; no rules – or at least that is how it seems. Hanging on by my fingernails as we hurtle into chaos; some hidden order prevents us colliding. Tuk-tuks weave and dart between us, two abreast in our lane alone. Lives hang in the balance. But I feel strangely calm. I can do nothing more than surrender my fate to my Mumbai taxi driver. And let myself just become a part of the flow of this place – the noise, the smell, the colour, the dust.

Nowhere is poverty and social inequity more extreme than here. And there are health problems that I have never seen in my work as a young doctor in the sheltered, distant world of coastal Western Australia. Only in my early work in remote Australian Aboriginal communities had I see anything quite like this, when, as a medical student with the Royal Flying Doctor Service, my job had been to swat away the flies as the doctor performed an emergency lumbar puncture (spinal tap) for an infant with bacterial meningitis on a dirt runway in the middle of the bush, with only an icebox as a makeshift bench.

I have only just arrived and already India is having an effect on me. It is two years since I received my PhD and I am looking for inspiration that will carry me forward into the rest of my career. I have recently been appointed jointly as a consultant paediatrician at the main children's hospital in Perth, and as a senior lecturer at the University of Western Australia (UWA). It is an exciting time, as I start to carve my identity as an independent researcher. As always, life is so busy that it is easy to forget to step back and look at the bigger picture. But here in India, miles from home, everything seems much clearer. It is time to think big and set some long-term goals. Amid a foreign world and among strangers I feel a new need to find 'a greater purpose and a greater happiness'.

I am here on a different kind of pilgrimage from the usual 'ashram' kind, if I survive the taxi journey. Research scientists, medical doctors, professors and assorted others are converging on India from all over the world to discuss the early-life origins of disease. I feel very excited to be part of this momentous venture. It is a major milestone for world health, even though many are yet to realise it. For the first time, this meeting will bring together experts from almost every field of medicine with a common goal: promoting the importance of a 'healthy start to life', beginning from (and even before) conception, as the best hope of preventing disease later in life. This 'cause' seems so obvious now that we might well question why it has taken so long to examine it like this. But we are now doing so, and that is something to celebrate.

The importance of a healthy start applies to virtually all body systems. Adverse conditions in early life can have lasting effects on all aspects of growth and development. These very early effects on both structure and function shape our physiological (functional), immune, metabolic, and even psychological and behavioural response patterns to the environment, and can have lifelong effects. Most importantly, these effects can influence our susceptibility to diseases decades later. It is logical that promoting optimal conditions in early life is the best hope we have of hardwiring 'healthy' physiological, structural, immune, metabolic and behavioural-response patterns in order to prevent so many avoidable diseases. Some of the best-known examples of this are how foetal growth patterns can influence the risk of heart disease, obesity and type 2 diabetes in later life. And there are many more examples. But more of that later.

I am in Mumbai as a representative of the 'immune system', which is also exquisitely vulnerable to the early environment and has a central role in the rising risk of many different diseases in the modern world. I have been studying the immune system for years, becoming a specialist in the field. With all specialists there is a danger that our focus can take us away from the bigger picture, and that our knowledge of other systems and important intercon-nections can lapse. I fear that has slowly become the case with me. Until now. This meeting is a real medical 'melting pot', and an enormous and exciting chance to learn about many other fields of medicine, all with a focus on early-life origins of health, but with their own unique perspectives. The time is ripe for discovering common ground. The most important and impressive achievement of the organisers has been the effort to be 'all inclusive' and to invite people from almost every field of health and medicine, as well as politicians, policy makers, and even religious leaders.

As my taxi brings me ever closer to the air-conditioned, five-star hotel, the challenges to both individual health and societal health are apparent at a single glance. Along each side, the highway is lined with makeshift housing, built from scavenged scraps of metal, wood and concrete. There is a woman washing in a ditch. Another preparing food over a smoking cooker. An old man performing his ablutions for the passing traffic to see. Bone-thin children playing on a pile of rubbish. Pollution, poverty, limited nutrition and the ever-constant risk of infectious diseases are the most obvious threats here. But this is a society in rapid transition, now faced with a very different set of problems: rising rates of obesity, heart disease, cancers, asthma, diabetes and all the other conditions that we associate with an affluent 'Western' lifestyle. Although this might once have been most obvious in the growing middle classes, the rate of these 'Western' diseases is now also rising more rapidly in the poorest classes. We now know that many of these diseases are preventable, and the best chance of reducing the rising global burden of these 'non-communicable diseases' (NCDs) is by understanding and addressing the modern lifestyle and environmental changes that are driving this. This issue is now just as important in the 'developing' regions of the world as it is in the more 'developed', industrialised regions. In fact, poverty is a major risk factor for coronary heart disease and many other NCDs, even in developed countries.

As we pull up to the Oberoi Towers by the water's edge, I am partly sad and partly guilty to leave the crazier side of Mumbai at the doorstep. At the next chance I get I will be back out on the streets, where I will find that a pale, blond, young woman walking alone gets far too much attention, even though I have been careful to cover my shoulders and dress with modesty.

The Oberoi lobby is abuzz with arriving delegates. Not only is this a world first, it is the official birth of a new health movement. There is an electric sense of anticipation. At the opening event, I am warmly welcomed by Dr Caroline Fall from the Southampton-based organising committee. She briefly introduces me to Professor Mark Hanson, who also thanks me enthusiastically for coming. Then I slip quietly into the crowd, hoping to catch a glimpse of the famed Professor David Barker. I don't actually know what he looks like yet, but everyone here has heard of the 'Barker hypothesis'. To a large extent, this is the reason that we are all here.

Like everyone else, I have read the acclaimed work that David Barker and his Southampton team published in *The Lancet*. In 1986, they first described geographic studies showing that the wide

4

regional variations in coronary heart disease throughout England and Wales might be linked to impaired foetal growth and nutrition in pregnancy.[1] Much higher rates of coronary heart disease were seen in poorer rural or industrial areas where infant mortality had been also been highest. They speculated that the in-utero risk factors for infant mortality in these regions were also risk factors for poor foetal growth, and that this poor early nutrition might be the contributing risk factor for later heart disease in the region. To really test this idea they had to find a very large group of people in middle age or older, whose birth weight had been accurately recorded. They found this opportunity in Hertfordshire, UK, where from 1911 onwards midwives and health visitors recorded birth weight and visited homes to measure infant weights. This formed the basis of their next *Lancet* paper[2] in 1989, which found that the risk of coronary heart disease in over 5,000 men born between 1911 and 1930 was proportionally higher as their birth weight decreased. This appeared to confirm their earlier theory. In the years that followed they also published research showing that adult lung function and chronic lung disease are similarly related to early nutrition, birth weight and other early-life events. As we will see later on, there is now conclusive evidence that patterns of growth and nutrition in early life can influence the risk of not only heart and lung disease, but a wide range of other NCDs.

These early Barker papers served to challenge the orthodox view that most disease resulted largely from adult lifestyles and genetic inheritance. There were actually even earlier reports, by Forsdahl in 1977,[3] that poor living conditions in childhood and adolescence might be an important risk factor for heart disease much later in life, but the significance of this was not appreciated at the time.

The concept of 'early development as a critical time for pro-gramming later risk' provided a new perspective and offered a new way forward. Prevention is the ultimate goal, and it is already clear that trying to prevent disease by only changing adult lifestyle is not very effective. A broader, longer-term vision is needed, starting at the very origin of life. For me it is a very exciting moment to see David Baker make his way onto the podium to tell this story.

So why is a paediatrician and immunologist with a focus on allergic disease so intrigued by ideas about the origins of adult heart disease? As it happens, my first research thesis investigated the effects of diet and lifestyle on blood pressure. In 1986 I worked with Professor Lawrie Beilin at the Royal Perth Hospital, carrying out a dietary intervention study in adults. I was following in the footsteps of Dr Barrie Margetts, who had done his PhD on a related cardiovascular topic. Barrie had been literally packing his bags, about to move to Southampton in the UK, when I visited him to get some pointers for my new project. I sat amongst the packing boxes to get his advice, which is where I also first heard about his plans to work with David Barker's group in Southampton. After he left, his name appeared as co-author alongside David Barker's on the aforementioned 1989 *Lancet* paper. It is not unexpected that he is here in India too, but it does make it seem like a small world.

As I walk over to say hello, I don't think Barrie recognises me. This is hardly a surprise, as when he met me more than fifteen years ago I had a black punk hairstyle. It is lovely to make the connection with him again, and to have the chance to bring him news of the Royal Perth Hospital team, with whom I still collaborate.

I inevitably also meet up with the other experts on the immune system. But for me the most interesting and important part of the conference is a chance to meet people studying other systems, and to hear lectures from experts in other fields. As expected, there is a heavy focus on the early-life risk for heart disease, obesity and diabetes. But other disciplines are well represented and, among these, I am most fascinated to learn about the developmental programming of brain and behaviour, and how very early experiences can mould the developing neural networks in ways I had never imagined. Soon it is my turn to give my presentation. And I hope that I have convinced a few people that the early programming of the immune system is more than a side issue.

Coming here, it is becoming even clearer to me that the immune system is a core part of the wider NCD story. My early research into heart disease taught me a lot about the role of inflammation in hypertension and cardiovascular disorders. In fact, it has become increasingly obvious that inflammation is a common element of most NCDs, including heart disease, lung disease, joint disease, bowel disease, diabetes and obesity-related diseases. Modern environmental changes appear to promote inflammation and, as a paediatrician, I have already seen the early-years effects of this in the epidemic of allergic inflammation. This is the clearest evidence we have that the early immune system is exquisitely sensitive to modern environmental changes. And I can't help wondering if the same environmental risk factors might be implicated in the increase in other inflammatory diseases. Diet, microbes, exercise and modern pollutants all affect the developing immune system, and these are also linked with the rise in so many other NCDs.

I had been invited here because of my own *Lancet* paper,[4] published two years earlier (1999), which turned the spotlight onto the 'early origins' of allergy and immune diseases. My work with Professor Patrick Holt in Perth was the first to map immune responses over the first years of life and to find early differences in children with allergic inflammation. Some of these differences were already apparent at birth, providing evidence of 'foetal programming' of allergic disease and fuelling interest in much earlier events. Differences in the very-early postnatal maturation of immune responses also indicated the importance of events in the first months of life. As with the Barker story, this contributed to a significant shift in thinking towards the early origins of allergic disease. In some ways this was an easier case to sell than the early origins of heart disease: because eczema, food allergy and asthma appear so much earlier in life it is more obvious that early events *must* be important.

And for this reason, my current research is focused on trying to prove that a number of specific early environmental factors, such as maternal diet, smoking and early microbial exposure, have at least some of their effects on the developing immune system (Chapter 10). This meeting in Mumbai has already left me committed to also examining the effects of the same exposures, in the same children, on other systems such as the developing brain, heart and metabolic responses. If we can find clearer evidence of connections, then it will be easier to convince people that we need to take a more interdisciplinary approach to disease prevention.

After my talk, there is an important chance to rediscover someone I already know and much admire Professor John Newnham. John is a fellow traveller from Perth, who had been a young academic in the Department of Obstetrics and Gynaecology at UWA back when I was a medical student. I came across John again recently after being appointed as a senior lecturer in Paediatrics and Child Health. Tasked with developing aspects of a new medical curriculum, I had also to liaise with John and his staff in the Department of Obstetrics and Gynaecology. But until I arrived in Mumbai I had not really had a full appreciation of the 'developmental origins of health' work that was already well underway in Perth under John's expert guidance.

John is associated with the Western Australian Pregnancy (Raine) Study, which began in 1989 and is the largest and best-known prospective pregnancy cohort in Western Australian history, and is also one of the world's very first truly multi-disciplinary research initiatives focusing on the 'developmental origins of health and disease' (DOHaD). It is also the first and largest study in the world to include detailed foetal ultrasound measurements, along with detailed long-term follow-up of the children after birth. The cohort is named after Mary Raine – a wealthy businesswoman who donated a fortune to help start a new medical school at UWA in the 1950s. As the vice-chancellor recruited for this purpose in 1953, my grandfather Sir Stanley Prescott helped broker the deal and set up what was to eventually become the Raine Foundation in 1957. Its vision had an enormous impact on medical research in our state, and hundreds of young researchers have had the benefit of this since.

Later that evening John also introduces me to Professor Karen Simmer from Adelaide, sings her praises and explains that we might be seeing more of her. She is a neonatologist whom they have recruited to UWA, and she will be joining his department in the coming months. We immediately find common ground. She is already well known for studying the benefits of omega-3 polyunsaturated fatty acids (PUFA) in fish oils for the developing brain. I have been studying the immune benefits of fish oils in pregnancy for allergy prevention. I can already see the opportunity to work with Karen to examine the neurodevelopmental effects in my allergy cohorts, glad that my new resolution to look at multi-system health benefits through new collaborations is already taking shape.

Each day I take the chance to go on a small adventure through the streets of Mumbai. To the parks where I see signs that exercise is 'strictly prohibited'. To beautiful temples; historic sites; a cricket match; shopping delights…

It is nearly time to leave, and I feel pleased that I have had my own life-changing experience. All this talk of being more healthy makes me realise that I have not been feeling very healthy. I leave India determined that 2002 will be a much better year.

• • •

And it was. In the year that followed I transformed my life and discovered a confidence that I did not know I had. Getting a promotion was the first step. The string of prestigious research grants helped too: all for 'early origins' work, of course.

Now the light was really shining. I lost weight. Got fit. Felt healthier, happier and more positive than ever before. I was in love with life again. Things went from strength to strength. I felt I was now well on my way to finding my greater happiness.

The reason for mentioning any of this is to stress that it is never too late to make changes in our lives. Although our genes and early-life programming have an important role to play, they are not the full story. The key focus of this book might be the 'early-life origins of health', but there is also no doubt that healthy lifestyle choices and healthy thinking can make a difference at any stage of life. This is why there has been a progressive focus on taking a 'life-course approach' to promoting health and preventing disease, stressing the importance of taking these opportunities *throughout* life. Our purpose for stressing 'early life' in particular is that benefits may be greatest in this time, and that this critical period has been overlooked for far too long. In doing so, the last thing we want to do is to leave people with any feeling that it is all too late for them to make a difference. For all of us, *the first and best opportunity we have for change is 'right now'!*

Professor Alistar Robertson from our vice-chancellery asked me what exactly were we trying to achieve in setting up our national Australian DOHaD society. A simple question. 'And don't be afraid to aim high', he said. So I thought about what we all really want, and said to him:

> *We want to achieve true physical, mental and*
> *spiritual well-being of individuals and of societies,*
> *and*
> *to maximise all opportunities for achieving this*
> *beginning with the earliest stages of life.*

And why stop there? When we consider it in those terms it is immediately apparent that the challenges we face go well beyond the health sector, and that we can only address this in the much broader context of the social, political, cultural and economic determinants of health. But as with everything, the first step is the intention and having the vision. And it needs to be a common, unifying vision.

2

Striving for a common global vision

It's almost ten years since the Mumbai conference, and a common, unifying global vision is finally emerging. A major global health crisis has been building, slowly and largely unheeded, for decades. The first step in solving any problem of such magnitude is to recognise it, acknowledge it and collectively agree that action is needed. Even this initial step can take considerable time and effort. We have to contend with sabotage from serious villains – Big Tobacco, an industry built on addicting youth and promoting chronic ill health and disease, and the Food-and-Drink industry, which contributes damaging levels of fat, sugar, salt and 'fast' food. While we could ban tobacco, we can't ban food, and a more challenging, collaborative approach with industry will be needed. The first step is to agree that we need to act.

But this most important first step has now been taken, and at last we seem to be crossing a great divide. There is now a clear, deep commitment at the highest levels of government, and a strong sense of possibility and optimism that we must not let pass. Though it may be a crisis of our making, like so many of our global challenges, we have the power to overcome it. But only if we stand together. If we don't, it will surely be our undoing.

A GLOBAL CRISIS – A GLOBAL RESPONSE:

In 2011, in response to this global health crisis, the United Nations (UN) General Assembly held its first high-level summit specifically to address the rising worldwide burden of non-communicable diseases (NCDs). This brought the issue squarely into the global political agenda, with the participation of heads of state and governments from around the world. The significance of this historic event cannot be underestimated. It is only the second time that the UN General Assembly has *ever* met with a health issue at the core of its agenda. And they met for good reason. Described by His Excellency Mr Ban Ki-moon, UN Secretary-General, as a 'public health emergency in slow motion',[1] NCDs are now seen as a major global threat to humanity, not only to our health, but to the social and economic advancement of all nations. These 'modern-lifestyle' diseases are now recognised as the leading cause of death worldwide, in *both* developed and developing countries. They are a major cause of poverty and must be viewed as a major global economic threat.

The statistics behind this are staggering. Based on the sobering figures presented in the UN *Political Declaration on the Prevention and Control of Non-communicable Diseases,* there are more than 36 million deaths each year from NCDs.[2] These chronic diseases account for more than 60 per cent of all global deaths. Tragically, at least 9 million of all preventable NCD deaths are in younger people, in their 'prime' years of life. An astounding 80 per cent of all NCD-related deaths occur in the developing, underprivileged regions of the globe.[3] Without urgent action it has been projected that NCDs will claim the lives of 52 million annually by 2030. Even in Africa, NCDs are projected to well surpass infectious diseases as the most common cause of death by 2030. No region, no population is unaffected.

So it seems that while the international community has been focusing on reducing the burden of infectious diseases (such as malaria, tuberculosis and HIV-AIDS) in the developing world, an epidemic of NCDs has emerged almost unnoticed in these regions. Specifically, this includes soaring rates of cardiovascular diseases, diabetes, cancers and chronic respiratory diseases. There is no doubt that this epidemic is hitting the developing world and lower income populations hardest. While poverty and lack of education are major risk factors for NCDs, NCDs also drive both individuals and economies further into poverty, creating a vicious cycle. It is clear that this cycle can only be broken by addressing the social and economic determinants of health, underscoring the compelling need for a multilateral 'whole-of-government' approach.

What really motivates politicians is economic cost. Lost workforce, lost productivity, and burgeoning healthcare costs. As described by Dr Margaret Chan, director-general of the World Health Organization (WHO), these are the diseases 'that break the bank'. Diabetes care is but one example where healthcare costs already account for 15 per cent of some national budgets. In the absence of urgent action, the rising financial and economic costs of NCDs will reach levels beyond the capacity of even the wealthiest countries. According to a 2011 report from the World Economic Forum, NCDs could cost the global economy more than $30 trillion over the next twenty years, equivalent to 48 per cent of the 2010 global gross domestic product (GDP).[4]

In a response proportional to this international crisis, the United Nations called its momentous meeting at the highest level, bringing together political leaders to develop a new global agenda: of reducing premature deaths from NCDs. By its very nature, this meeting took the critical step of promoting a whole-of-government approach to preventing and controlling these diseases.

The United Nations political declaration presented before the General Assembly proposed that 'prevention must be the corner-stone of the global response to NCDs'.[5] The resulting dialogue between the political, non-governmental and business sectors has led to unprecedented consensus. And it could mark a major turning point: a real opportunity to make major advances.

BRINGING 'EARLY-LIFE ORIGINS OF HEALTH' TO THE TABLE:
Prevention is clearly the name of this game. It is stamped all over the global strategic plan. If prevention is to be more effective, it is obvious that strategies need to be implemented *before* disease processes are initiated and *before* risk factors come into play. With clear evidence that the pathways and risk factors that lead to many NCDs begin to have effects very early in development, this should be a core element in the agenda and strategic plan. But it has not been.

The challenge was to bring this philosophy to the United Nations agenda with the hope of incorporating it into the strategic core.

I was privileged to play a very tiny part in an effort to address this need, through the World Universities Network (WUN). Both the University of Southampton and my university (UWA) are members of the WUN. The mandate of the WUN is to address a number of key global challenges through collaborative research. This WUN Global Challenge Program addresses major international issues such as 'Adapting to Climate Change', 'Understanding Cultures' and 'Globalization of Higher Education and Research'. WUN recently also established a 'Global Health Challenge'. With Southampton's strategic interest and leadership in Developmental Origins of Health and Disease (DOHaD), there has been a strong emphasis on 'early-life opportunities for the prevention of NCD in developing countries'.

Largely engineered by Professor Mark Hanson and Professor Sir Peter Gluckman, the first meeting on the topic was held in Shanghai, China in May 2011, strategically ahead of the United

Nations General Assembly meeting by several months. I was invited to attend on behalf of UWA and so was John Newnham.

Following the success of the Mumbai meeting, the DOHaD Society had continued to hold international meetings approximately every two years. In 2007 we hosted the meeting in Perth, and John was the conference chairperson. John had continued to play a key role on the Raine Study (the Western Australian pregnancy study). By 2011, the Raine Study Group already comprised more than 150 senior researchers providing expertise across a wide range of fields, with many more junior staff and PhD students in tow. They had generated more than $25 million to follow thousands of the study's children into adulthood, to better understand the effects of early-life factors on all aspects of subsequent health. This truly interdisciplinary study team has become internationally recognised for leading one of the most detailed birth cohorts in the world, with more than 160 prominent research publications since it first began in 1989.

Over that same ten years since Mumbai I had been progressively developing my own research group. My team focused on cardiovascular and metabolic measures; speech, behaviour and neurodevelopment; and my speciality, allergy and immune development in early childhood. In 2010 this effort was recognised in a national award known as '10 of the Best' by the Australian National Health and Medical Research Council. My international profile in allergy and immunology had been building strongly, and I had been using this to promote the DOHaD agenda in my own discipline at the highest levels.

Over this period I had also continued my work in a busy paediatric allergy clinic, watching our waiting lists becoming longer every year. For my patients' parents, hungry for answers, I wrote my first book, *The Allergy Epidemic, a Mystery of Modern Life.*[6] I discovered that my clinical, research and international activities had positioned me well to start a useful dialogue with the public on these wider issues, and to make a new contribution that way. After finishing my book, my husband joked that he was worried about what I might do next. I do like a challenge and I *was* looking for a new direction.

Arriving in Shanghai in 2011, I could feel an opportunity for just that.

The Shanghai agenda included understanding the drivers of risk (the most important early-life determinants of risk for NCDs); what can be done, including early-life intervention to be facilitated by partnerships with governments, industry, media, agencies and non-government organisations (NGOs); and, finally, wider social, economic and political issues, particularly in the developing world, and the best avenues to influence policy.

But the most important task of all was to develop a simple, clear and unequivocal declaration to be delivered to the United Nations, ahead of the General Assembly.

By the end of the meeting we had produced what has become known as the *Shanghai Declaration*,[7] led by Mark Hanson and Sir Peter Gluckman, with input and endorsement from the other delegates including leading academics from non-WUN universities in Latin America, Asia, the Caribbean and Africa.

This resulting document stressed that initiatives to promote a healthy start to life can reduce the risk of later NCDs, and that these interventions should be part of an integrated life-course strategy to reduce the burden of NCDs, alongside proposed adult interventions and treatments.

Such an approach highlights the central importance of health literacy and, in particular, the need to educate and empower women of child-bearing age in all regions of the world. The *Shanghai Declaration* stressed the social and economic benefits of low-cost interventions applied *early* in the 'life course', and the wider benefits of a healthy start to life, education and gender equity. It also endorsed a multi-agency approach linked to current programs on maternal and child health and infectious disease and to the Millennium Development Goals.

It was pretty exciting to be part of this. I was asked to speak on *lessons from early programming of immune function*. I discussed the fact that more than 300 million people worldwide have asthma, and many others have eczema and other allergic conditions which affect life quality and negatively impact the socio-economic welfare of society.[8] I pointed out that the epidemic rise in allergic conditions (now affecting 30 to 40 per cent of the population) also suggests similar risk factors as for other NCDs. Allergic diseases appear *much earlier* in life than other NCDs, suggesting that the developing immune system is particularly sensitive to environmental changes. By extension, this means that the immune system is a useful *early barometer* of environmental impact, and an early measure of effectiveness of any interventions that we might try to prevent disease.

It was an important, albeit small, step forward to see allergic diseases added to the list of NCDs noted in the footnotes of the *Shanghai Declaration*.

After the meeting, the *Shanghai Declaration* and consensus position was published in *The Lancet* on 13 August 2011 by Mark Hanson and Sir Peter Gluckman, along with other correspondence, ahead of the UN General Assembly. This correspondence was co-authored by Professor Don Nutbeam, vice-chancellor of the University of Southampton, who has experience with health literacy, and by Professor John Hearn of Sydney University and chief executive of the WUN. As originally intended, the *Shanghai Declaration* was also submitted to the UN at the request of Dr Ala Alwan, Assistant Director General of WHO. Then we all had to sit back and see if it had any impact on the UN agenda.

THE FIRST 1,000 DAYS:

In the *UN Political Declaration on the Control and Prevention of Non-communicable Diseases* there was a small clause in which the Assembly

'*Note with concern that maternal and child health is inextricably linked with NCDs and their risk factors, specifically as prenatal malnutrition and low birth weight create a predisposition to obesity, high blood pressure, heart disease and diabetes in later life; and that pregnancy conditions, such as maternal obesity and gestational diabetes, are associated with similar risks in both the mother and her offspring*'.[9] There was also some mention of the need for including NCD prevention and control 'within sexual and reproductive health and maternal and child health programs'. But that was pretty much it.

I think we can be happy that *some* progress was made, but it is clear that a lot more is needed.

Prevention must be the ultimate approach, and the enormous weight of evidence and logic is that this *must* be directed as *early* in life as possible, when all structures are formed, physiologies established and behaviours patterned. This must be a *core and central* platform in strategies for change, rather than a noted side issue. And we must all continue to prevail on leaders at every level, and in every sector, to buy in to this long-range, lower-cost vision that is ultimately likely to be more effective.

On a far more positive note, the importance of early-life programming has gained considerable traction elsewhere. In September 2010 US Secretary of State Hillary Clinton launched the '1,000 Days' movement, bringing together numerous international organisations with a shared purpose of promoting optimal growth and nutrition from conception for ensuring future health. Although this movement is focused specifically on nutrition, it recognises the first 1,000 days (from pregnancy until two years of age) as a critical window of opportunity to achieve measurable and lasting impact on a child's development by enhancing intellectual, physical and social growth. With political support at the highest levels, this is now firmly on the international agenda.

In the words of Hillary Clinton, 'we believe fervently that improving nutrition for pregnant women and children under two is one of the smartest investments we or anyone can make'.[10] The movement enlists other agencies to support its agenda through agriculture, education, employment, social welfare and development programs which aim to empower women to improve their own and their children's nutrition. Overall coordination of the 1,000 Days movement is being led by a large coalition of NGOs (InterAction) in partnership with the United Nations Scaling up Nutrition (SUN) program, which brings over 100 organisations and governments together to fight hunger and under-nutrition.

This is a clear and decisive response to the Millennium Development Goals (MDG) set for 2015 of 'Reducing poverty, hunger and under-nutrition' (MDG1), 'Reducing child mortality' (MDG4) and 'Improving maternal health' (MDG5). The United Nations is heavily invested in this program, and UN Secretary General Ban Ki-moon has continued to express his deep commitment to this endeavour, *'We must end the hidden tragedy of stunting, which affects 200 million children. Food and nutrition security are high on my action agenda for the next five years. I urge all partners to do their utmost to rise to this challenge. Together, we can unlock the potential of current and future generations.'*

Poor nutrition is a major risk factor for both infectious diseases and NCDs, and is of critical importance in the first 1,000 days. At the same time it is equally important that any endeavour to promote optimal conditions in pregnancy and early childhood *also* addresses other risk factors, in particular avoiding adverse exposures in this period.

The most obvious example is cigarette smoking. This is still one of the most devastating public-health challenges in both developing and developed countries. A tobacco-free future is the ultimate goal of the United Nations *Political Declaration for the Control and Prevention of Non-communicable Diseases*.[11] Political consensus may have been achieved, but Dr Margaret Chan, director general of WHO, has continued her call for heads of state and government to 'stand rock hard' against the highly aggressive 'despicable' tactics of the tobacco industry for real progress to occur.[12] Exposure to cigarette smoke in the first 1,000 days is a particularly critical risk, and it is vital to include a specific focus on tobacco control in teenage girls and young women of child-bearing age as part of this broader agenda.

A less well-known issue is high levels of toxic indoor pollution from dirty stoves and fires, which kills almost 2 million people each year, most of them women and children. It is estimated that 3 billion people live in homes where food is prepared on stoves or over fires that produce toxic fumes and chemicals at hundreds of times the levels considered safe to breathe. The on toll on human health at every age is considerable (including cancer, pneumonia, cataracts, low birth weight and even death), but with particular implications for women (who prepare the food) and their children in the first 1,000 days. There is also now a *Global Alliance for Clean Cookstoves* lead by the United Nations Foundations in partnerships with governments, not-for-profit organisations and the private sector.

While the NCD agenda is largely focused on poor nutrition, tobacco, alcohol and lack of exercise as the big four risk factors for chronic disease, it is important to recognise that the environmental determinants of modern diseases are likely to be far more extensive and more complex than that.

In addition, we must recognise the complex interrelation-ships between infectious or 'communicable' diseases and NCDs. Infections in infancy can be an immediate threat to life, and the broad-ranging strategies to reduce these are a major imperative, particularly in developing regions of Africa where they still outstrip NCDs. At the same time, we have also seen there may be a price to pay with progressively more hygienic environments (Chapter 5). In highly developed societies, changing microbial exposures have been linked with a number of NCDs including asthma, allergies and other immune diseases, diabetes, cardiovascular disease, obesity and even behavioural and psychological disorders.

We need to apply these lessons to anticipate, curtail or even prevent the same adverse consequences in developing regions as they undergo rapid transition to 'Western ways'. We should be urgently trying to understand the much wider consequences of Westernisation, beyond the 'big four' risk factors. We can only hope that it is not too late.

THE PARADOX OF INEQUITY AND TRAGEDY OF PLENTY:

The dominant focus of the '1,000 Days' program is on nutrition and in particular *under*-nutrition. With over 1 billion people in the world today still under-nourished this is an undisputed and critical priority. But 'the other half' of the world's population is suffering growing ill health and chronic disease because of *over*-nutrition. *Both* increase the risk of NCDs.

There's often an overwhelming sense that it is all too hard.

It isn't. This is an age of unparalleled technology and scientific advances. And our discoveries, our networks and communications are all growing at lightning speed with every day. Despite the global financial crisis, the world has never before seen such prosperity.

And for the first time our leaders and whole governments are behind global initiatives to address this inequity, through programs such as the Millennium Development Goals (below). These also have the support of major international financial institutions such as the World Bank, the International Monetary Fund (IMF), the regional development banks, and members of the World Trade Organization. We have the best chance of any generation yet. And we must all take ownership of these goals and hold our leaders to their commitments.

THE MILLENNIUM DEVELOPMENT GOALS: HELP OR HINDRANCE FOR NCDs?

The *Millennium Declaration* was signed at the United Nations Summit in 2000, as a historic promise by 189 heads of state and government to end world poverty by 2015. The *Millennium Development Goals* (MDG) provide the eight-point 'road map' intended to achieve this. An associated global public campaign (the UN Millennium Campaign) was also established to inspire people to sign up and get involved as 'the generation that puts an end to poverty', and to hold governments to their targets.

In essence, the first seven goals were targets for the poorest countries of the world to work towards by 2015, with measurable targets and clear deadlines. The eighth was to achieve partnerships that ensure accountability and efficient use of resources, and to have richer nations support this process through effective aid, sustainable debt relief and fair trade.

- *The first goal: reducing poverty, hunger and poor nutrition.* For example, the number of people living in poverty (on less than $1 per day) and those suffering hunger should be halved.
- *The second goal: universal education.* All children, both girls and boys, to have access to primary education by 2015.
- *The third goal: gender equity and empowerment of women.* Gender disparity should be eliminated in education by 2015.
- *The fourth goal: reducing child mortality.* By 2015 mortality for under-fives should be reduced by two thirds, particularly from pneumonia, diarrhoea, and measles.
- *The fifth goal: reducing maternal mortality.* Maternal mortality should be reduced by three quarters before 2015, with improved access to antenatal care and contraception.
- *The sixth goal: reducing HIV/AIDS and other infectious diseases.* By 2015 the spread of HIV should have begun to be reversed, and treatment should be universally available. Other target diseases include tuberculosis and malaria.

- *The seventh goal: environmental sustainability.* Improved sanitation and water supply: by 2020, the goal is for significant improvement for at least 100 million slum dwellers. It also addresses the loss of biodiversity and climate change.
- *The eighth goal: global partnership.* Poor countries will work towards achieving the Millennium Development Goals, with greater accountability to citizens and efficient use of resources. Rich countries will deliver more effective aid, and fairer trade, including benefits of communication technology, well before 2015.

It was recognised that health is intrinsic to development. Targets for child health, maternal health and infectious diseases are health-related at their core. And targets for poverty, hunger, water, sanitation, education and empowerment of women all address the wider social and environmental determinants of health.

However, the landscape has changed substantially since the year 2000 when this vision was developed. The growing crisis of NCDs was not recognised in the developing world at that time. The unanticipated social and economic burden of NCDs has become a serious impediment in achieving virtually all of the MDG targets, and in particular Goals 4 and 5 (children's health and women's health). Ironically, it is these goals that are likely to be most important in breaking the cycle and reducing the future disease burden; yet it is these goals which are falling the furthest behind.

Having specific targets provides focus and accountability and allows funding to be channelled appropriately. However, because NCDs were not explicitly mentioned, many governments and donors have not funded them as a matter of policy. More than ten years into the Millennium Program, the United Nations has taken serious measures to address this through the 2011 meeting of the General Assembly on NCDs. The wheels turn slowly, but this new focus on NCDs at the very highest levels will hopefully help achieve the shift in perspective that is needed down the funding food chain. Saving billions of people from infection without rescuing them from the more insidious epidemic of NCDs that is now threatening these populations will achieve little. A prime example is the vast investment in fighting the millions of deaths each year from tuberculosis, without specific targets for a more common cause of chronic lung disease, smoking, which kills almost twice as many.

Another example is diabetes. Although more than 80 per cent of all people with diabetes live in low- or middle-income countries, diabetes was not considered a development priority. In India alone there are an estimated 50 million people living with diabetes, and in some small pacific island states diabetes affects almost one-third of the population. Indigenous populations with a greater genetic predisposition, such as Australian Aboriginals, are particularly vulnerable to diabetes. Diabetes is still a neglected cause of maternal mortality, birth complications and infant mortality. Compounding this, infants of diabetic mothers have a much higher rate of diabetes in later life. And yet, diabetes was not included, and therefore not funded, in the development goals.

It is vital that these are not seen as competing interests, and that the critical connections between NCDs and the existing Millennium Goals are better recognised. The only solution is sustained funding for NCDs in harmony with other goals and challenges.

The formation of the NCD Alliance was a critical step in promoting this agenda and lobbying for change. This alliance was founded by four international NGO federations representing four main NCDs: cardiovascular disease (World Heart Federation), diabetes (International Diabetes Federation), cancer (Union for International Cancer Control), and chronic respiratory disease (International Union against Tuberculosis and Lung Disease). Their mission is to put NCDs on the global agenda. In May 2009, the

NCD Alliance launched the successful campaign for a United Nations High-level Summit on NCDs.

Considerable progress has been made since, with the 2011 UN declaration on NCDs. But ensuring a focus on the developmental origins of NCDs and the specific early-life opportunities that address these is a battle we have yet to win.

BROADENING THE NCD FOCUS, INCLUDING IMMUNE DISEASE IN INFANTS:

With the NCD Alliance largely focused on what are described as 'the big four NCDs' and 'the big four risk factors', it is important that other extremely common NCDs and their risk factors are progressively incorporated into the agenda in a complementary, cooperative way.

At the moment the 'big four' diseases on the agenda are cancer, diabetes, heart disease and chronic obstructive lung disease. But there are other NCDs rising with environmental transition and modern lifestyles.

These include rising rates of mental illness and dementia, kidney and liver disease, arthritis and osteoporosis, allergic diseases, immune diseases such as thyroid disease, multiple sclerosis and inflammatory bowel diseases. And this is by no means the end of the list. Importantly, these conditions have an early-life origin dimension (Chapter 3).

Accordingly, the recent United Nations *Political Declaration on the Control and Prevention of Non-communicable Diseases* extended their agenda by recognising mental and neurological disorders, such as Alzheimer's disease, as part of the global NCD burden. It also recognised kidney, oral and eye disease. Even trauma and injuries from traffic accidents have been included in the NCD discussions. But there are a still even more NCDs that should be brought into this arena.

My own personal area of focus provides a perspective that may influence the way we see the larger picture, and is likely to also provide a few new clues.

Allergic diseases such as asthma, eczema, allergic rhinitis and food allergy are rapidly rising in almost all developing regions including China, India, South America, and Africa. These immune-mediated diseases are arguably the most common NCDs of all, particularly in younger people. Although allergic diseases are viewed as diseases of affluence, there is actually a socio-economic gradient, with more asthma deaths in less-affluent sectors of high-income countries, who also have more limited access to care. Lower socio-economic populations bear a disproportionate burden of exposure to pollution, suboptimal diet, and unhealthy environmental conditions (physical, social, and psychological). Again, conditions traditionally linked with affluence have been rising almost unnoticed in the poorest regions.

Another interesting parallel with other NCDs is the differences in genetic susceptibility between races. We already know that non-Caucasians are more susceptible to NCDs such as diabetes, and there is now clear evidence that the same is true of allergic diseases. For example, in countries such as Australia, allergy rates are much higher in Chinese people than Caucasians, and the rates are proportional to time spent living in Australia. We are already seeing the allergy epidemic striking in Asia, and this means that means that we can anticipate an even greater global burden in the coming years.

The epidemic of allergic disease seems to be a very large signpost, pointing us straight to the immune system. Allergies are the only common NCD to appear well before the first birthday, with children bearing the brunt of the very dramatic recent rise in disease. The immune system is *designed* to respond to the environment. And quickly! That is how we survive. This may also be the

27

reason that modern environment changes lead to symptoms of immune disease so early in life.

Later we will explore how many environmental risk factors influence immune function. We will see how changes in diet can specifically alter immune function in the gut with implications not just for immune diseases, but also for diabetes, heart disease, obesity and even neurological disorders. Could the immune system be a common link between many of the NCDs? Are environmental risk factors mediating their effects (at least in part) through immune effects? Is this the reason for the shared genetic propensity for many NCDs in some populations? Could this be due to genes in immune pathways that convey greater susceptibility to environmental exposures in some races? Indeed, chronic low-grade inflammation is a common feature of virtually all NCDs, indicating specific vulnerability of immune pathways to modern environmental changes. Reducing the risk of inflammatory responses through lifestyle and environmental interventions is likely to have benefits for the risk and progression of many other NCDs. There is a very strong case for promoting immune health as an integral part of NCD prevention.

• • •

And so to summarise. The 2011 United Nations *Declaration on the Control and Prevention of Non-communicable Diseases* has been of pivotal importance in recognising that 'the global burden and threat of NCDs constitutes one of the major challenges for development in the twenty-first century, which undermines the social and economic development throughout the world, and threatens the achievement of internationally agreed development goals'.[13] It has also been fundamental for making real progress towards a unified vision for health, including firm measures towards addressing the major risk factors for NCDs. However, while it is recognised that a large proportion of NCDs are inherently preventable, the prevailing approaches focusing on adult interventions are far less effective than early-life approaches. While the vision of promoting heath in early life is more logical in the long term, it is still too commonly overlooked. It is now essential that we promote a greater awareness of the importance of a 'life-course' approach for disease prevention, beginning with strategies to promote maternal health, both before and after conception.

Many things we need can wait,
children can not.
NOW is the time,
their bones are being modeled,
their blood is being made,
their brains and minds are developing.

To him, we can not say tomorrow,
his name is TODAY

Gabriela Mistral (1889–1957)
Poet, Nobel Laureate

3

Early life:
A critical time of risk and a critical
time of opportunity

The moment of our conception seems as mystical and miraculous as the Big Bang itself. In an instant, new life is made. It sets in motion a frenzy of clockwork molecular activity as DNA is unfurled and the building blocks of life are made. Precise and certain. Of unimaginable complexity. Our bodies come slowly into being and so our time on Earth begins.

In those moments all of our potential hangs imprinted in our genes, which tell of all that we might become. But it is a long road to maturity, and much can happen along the way. At every turn there is the chance that adversity might steer us from our predestined course. We are also programmed for resilience and have an inbuilt capacity to adapt to the different conditions we might encounter. These sometimes-subtle shifts in our structure, physiology and even our behaviour are driven by a primordial need for self-preservation. If our pregnant mother is struggling to find enough food, our foetal metabolism adapts to conserve energy. If she is stressed by her situation, our stress response systems are engaged to face a hostile environment. If there is infection, and it is not severe enough to terminate the pregnancy, our foetal immune system evokes responses that can protect us from the same pathogens after birth. These and many other responses are designed to prepare us better for what lies ahead, to aid in our immediate survival and our chances of reaching reproductive age so that we can perpetuate the life of our species. Once we enter middle age, and the biological pressure to reproduce has passed, evolution is much less geared to our longevity and survival. This may be why some adaptive responses, designed to help us through early life, can actually be detrimental much later in life. Evolution does not seem to care that the most creative and productive years of human life can be after the age of forty.

THE FOUNDATIONS OF LIFE:
During the months before our birth most of our major structures and body functions are put in place. Small or subtle happenings in this period can have slow ripple effects that may not be revealed for many years. Key organs, such as the brain, the heart, the kidneys and the lungs are all formed in this period. In many cases the full quota of cells in these organs is fixed at birth (in the heart and kidney) or soon after (the brain and lungs). Even unnoticed, early adverse events might reduce the quota of heart-muscle cells, or the number of functional kidney units (nephrons), or the lung capacity that we are born with. And much of that is set for life. Once these organs are formed we can't grow new heart muscle or nephrons, although stem-cell research is trying hard to overcome these biological limitations. Initially, this may have no obvious consequences, and we may be perfectly healthy. Then we are challenged and our reserve capacity proves more limited. If we have smaller lungs because our mother smoked in pregnancy, we will be more susceptible to chest infections and wheezing illnesses in childhood. Even in adulthood, our lung capacity will remain reduced, and we will be more susceptible to chronic lung disease much later in life. If we have smaller kidneys, we may not experience problems until late in life when we become more susceptible

31

to higher blood pressure and kidney failure. Again, this is because we have reduced 'reserve' and cope less well in the face of a normal age-related decline in function. The same analogies apply to many other organ systems.

Just like organ systems, the number of fat cells, or 'adipocytes', in our bodies is also set during early life, usually by the end of adolescence. If we have a larger numbers of fat cells by then, we have a much greater risk of obesity later in life, and all the diseases that go with that. These are one kind of cell that we all seem to want to lose! But unfortunately once they are there, they are there to stay, unless surgically removed. When we lose weight the cells just shrink in size. And they soon 'get hungry' and produce hormonal changes that make us want to eat more. This is one of the reasons it is a constant battle to lose weight and to maintain weight loss, made even harder if we had more fat cells to begin with. Here too there is very good evidence that factors in early life program our fat storage and the metabolic pathways that lead us to conserve our energy as fat (explored more fully in Chapter 4). The same early events also predispose us to the many chronic diseases associated with obesity, such as type 2 diabetes, fatty-liver disease, heart disease, stroke, joint diseases, dementia and some cancers.

Even the brain, which is remarkably resilient, may show delayed effects of early events (Chapter 7). Subtle variations in the early environment may manifest as altered susceptibility to behavioural and psychological problems in early life, or degenerative disease in later life. There is now evidence that inflammation in pregnancy, caused by infection or other events, can increase the risk of schizophrenia or other mental illnesses much later in life. Obviously, more significant early insults to the brain will have more profound effects, such as those seen with foetal alcohol syndrome or cerebral palsy, although in most cases of cerebral palsy the early insult cannot be identified.

In this context is worth noting that the developing brain shows greater 'neuroplasticity', or ability to recover. Although a young individual still cannot develop new brain cells, they seem to be better at developing new connections. It is possible for young children who experience massive brain injuries to recover almost normal function.

When I was working in the children's intensive care unit I saw several young children with major head trauma after car accidents. Their injuries were so bad that there was initially little hope even for their survival. Years after one of these cases, when I moved to work in another section of the hospital, I was surprised by a young man I did not recognise. He was visiting the hospital with this mother, and I recognised *her* immediately. She took great pride in telling me that her son was just about to start his first job. He walked with a slight limp but had surprisingly little effects considering the scale of his original injuries. It had been hard work. Years of rehabilitation; and all worth it. It is much harder for adults to 'rewire' their brains. An adult with similar injuries would not usually fare as well, but some recovery is possible and there are 'miracle' adult cases where all or some function is restored. So while early insults can have lasting effects, there is an inherent plasticity and resilience in developing systems during early life.

This early plasticity is core to our adaptability and survival. However, some adverse exposures, such as maternal stress or dietary deprivation in pregnancy, induce 'adaptive' changes in the foetus that favour short-term survival, but have long-term costs. As we will see in later chapters, these early adaptations may lead to hormonal and metabolic responses that can increase the risk of a number of chronic diseases in later life.

And so the foundations of our life are laid. Sir Peter Gluckman and Mark Hanson, great leaders in this field, have given us the perfect analogy:

'It is like building a house. If the foundations are not properly laid, no matter what is done after that, problems will emerge sooner or later. A subtle defect in the composition of the concrete used in the foundations may not matter at the beginning but in time it will start to crumble and if there is an earth quake the consequences will be much worse'.[1]

SOONER OR LATER:

Adverse conditions in early life can manifest their effects at many stages of life. The link is most obvious when the consequences are evident very early. We see this with certain birth defects, and a range of congenital and infant conditions that result from a specific environmental exposure (or deficiency) during critical stages of development.

The most devastating examples of environmental *deficiencies* that lead to early but lasting adverse effects include conditions such as spina bifida (from folate deficiency in very early pregnancy), rickets (from severe vitamin D deficiency in early childhood), and 'cretinism' (intellectual disability caused by iodine deficiency and significant maternal thyroid deficiency in pregnancy). It is because these occur when systems are developing that the effects are so significant. It is all a matter of 'timing'.

The same applies for the many *adverse toxic exposures* that show very early effects. Some of the best-known examples include the tragic effects of thalidomide and congenital rubella. Thalidomide was once prescribed to pregnant women to reduce 'morning sickness'. It was effective and well tolerated by adults. But in the developing foetus, the drug inhibits proteins important in limb formation, with horrendous effects. Within a few years of its introduction in the late 1950s, more than 10,000 babies were born with grossly malformed limbs (referred to as phocomelia or dysmelia). And it took several years before the link to the drug was made by Australian obstetrician William McBride and the

German paediatrician Widukind Lenz. This tragedy led to much stricter testing and regulation for drug licensing.

Similarly, rubella (a relatively mild infection) in the first trimester of pregnancy results in a 50 per cent chance that the baby will have severe birth defects, including heart defects, eye defects and blindness, mental retardation and abnormalities of other organ systems, known as 'Congenital Rubella Syndrome'.

These early examples also illustrate how relatively simple public-health measures can successful avert these conditions. Vaccinating teenage girls for rubella has successfully prevented countless cases of Congenital Rubella Syndrome. Newborn screening programs can detect thyroid and other metabolic disorders very early, so that treatment can begin before brain damage becomes irreversible. Adding folate to staple foods and recommending supplements for women planning pregnancy has reduced the risk of neural-tube defects (spina bifida). And there are many other examples of how early interventions have early-life benefits, with long-term implications. The personal, societal and financial benefits of these strategies are undeniable.

It's far more difficult to see relationships when the cause and the effect are separated by decades, and even across generations in some cases. This has been the great challenge of the DOHaD (Developmental Origins of Health and Disease) movement. The later effects, such as how foetal growth patterns might relate to adult-onset diabetes, obesity or heart disease, are harder to understand and harder to investigate.

The new field of epigenetics (Chapter 11) has given a strong biological foundation for understanding how these delayed effects might occur, and how environmental changes can alter our patterns of gene expression and future disease susceptibility. These discoveries have provided a critical lynchpin and another layer of credence to the DOHaD 'early-origins' field.

A NEW AND CHANGING LANDSCAPE:

In the last fifty years there has been a dramatic surge in knowledge and advances in foetal and newborn medicine. This rapidly changing terrain of 'early life' has substantially changed perspectives and generated a new set of medical, social and ethical challenges.

Before modern imaging, methods of assessing foetal well-being were fairly crude, such as listening for the heartbeat, detecting foetal movements and palpating the size of the uterus to infer foetal growth and condition. There was no real understanding of the processes or plasticity of development, or the true degree of maternal and early-environmental influences.

Now, sophisticated scanning technology allows us to safely see three-dimensional moving images of the foetus. It is possible to diagnose and even treat some conditions many months before birth. In the 1980s, pioneer surgeons began performing foetal surgery to treat heart defects, kidney defects and other congenital malformations. Keyhole surgical techniques are now minimally invasive and much safer.

Another recent change is how early the foetus can survive *outside* the womb. Survival rates of premature infants continue to improve, particularly those born as early as 23 to 25 weeks of gestation. Major advances in neonatal intensive-care technology have had a significant effect on these survival rates. A single course of antenatal corticosteroids in women experiencing threatened preterm labour has also significantly improved outcomes if her baby is then born preterm. The general health outcomes for these survivors have now also improved significantly, with much lower rates of long-term disability. Preterm care is increasingly important in the face of rising rates of preterm labour in *both* developed and developing world regions. Rates in developed regions such as the USA reached 12 per cent in the 2010.[2] Women in Western Australia are almost three times more likely to give birth to preterm babies compared to women in China. Interestingly, in this study John

Newnham and his collaborators in China have also shown a rising rate of preterm birth in Chinese women as they moved to more Western environments, suggesting that behaviour and lifestyle may be significant factors associated with preterm birth. While this is still poorly understood, it has significant implication for growth and development and the risk of subsequent NCDs.

We must also consider the impact of declining fertility rates and the fact that significantly more babies are now conceived by assisted reproductive technology (ART) such as in-vitro fertilisation (IVF). This actually raises several questions. First, we need to understand how and why modern environmental changes are reducing fertility, and if the same environmental factors are having adverse effects on the babies that *are* naturally conceived. Women are now much older when they conceive, and this has additional implications for foetal development. The higher risk of birth defects, Down Syndrome and other chromosomal defects is well recognised, and progressively increases in mothers over thirty-five years of age. But there are also potentially more subtle effects that might affect foetal growth and increase the longer-term risk of heart disease, diabetes and other later-onset conditions in their offspring. Second, we need to understand the effects of reproductive technologies on early development. The very early period of 'embryogenesis' is highly sensitive to both maternal and environmental factors. During these early stages after fertilisation there are very dynamic changes, as large numbers of genes are switched on and off to generate the many different tissue types and organ systems. This carefully orchestrated program of development is tightly controlled by epigenetic processes that are sensitive to environmental factors. In other words, subtle alterations in the biochemical and biophysical conditions at conception and during early embryonic life can influence the early patterns of gene expression (see 'epigenetics' – Chapter 11). While this is true whether conception occurs naturally or by ART, there is emerging evidence that the artificial in-vitro conditions may influence these epigenetic processes to produce both short-term and long-term effects on development and health. Compared with natural conception, there do appear to be higher rates of congenital defects and chromosomal anomalies.[3] It is important to stress that the vast majority of children conceived through ART are born healthy and without any birth defects. However, it is the as-yet unknown long-term effects that early, more-subtle epigenetic changes might have on disease risk later in the ageing process that need to be investigated further. Long-term studies are underway to investigate consequences, if any, in later life.

IT IS NEVER TOO LATE:
Although the benefits might be greater if we make healthier choices from a younger age, switching to a healthier diet, taking more physical activity, giving up smoking and reducing stress will have clear health benefits at *any age*. We must promote health strategies over the *whole* 'life course', recognising that 'earlier is better' and that 'early windows of opportunity' extend well into childhood and adolescence. Indeed, as the next generation of parents, strategies to improve the health and behaviours of older children and adolescents will not only benefit their own long-term health, but also their children's to come.

History gives us the clearest evidence that we can alter our fate. We have seen dramatic improvements in life expectancy over the last century, particularly in developed countries. Once, if people survived the perils of childhood, they lived into their fifties or sixties on average, and this was true even up until the early twentieth-century. Even today, there are wide disparities in life expectancy around the world. In 2012 the estimated average life expectancy from birth reached eighty years of age in Britain compared with sixty-seven years in India and less than fifty-five years in many African countries.[4] The contribution of improved conditions in early life and adolescence is often underestimated in this. Poor early-life conditions remain a significant factor in both these discrepancies and in closing the gap.

So, when we ask if we are better off than we were in the Middle Ages, the answer is 'almost certainly', at least if we live in a developed nation. But when we consider whether we are currently living to our full potential, the answer is 'almost certainly not'. Some may be, but the many millions are still dying prematurely from avoidable, chronic diseases. And a good proportion of these deaths are the result of unhealthy lifestyles, behaviours and environmental exposures, some beyond our individual control. Now, for the first time, children in the most affluent countries are predicted to have shorter life expectancy than their parents, simply because of obesity and the chronic diseases that result from that (Chapter 4).

If at least some of this is determined when we are too young to make informed choices, each generation has some dependence on those who come before. This is not about blaming our parents though. Many choices, opportunities, knowledge and understanding are determined by the much *wider* social, cultural and economic and environmental context. This means that we must take *collective* responsibility for the health crisis we currently face.

Our clear goals should be to provide a clean and healthy environment, the best knowledge and the best opportunities at every age, in particular for every prospective parent.

The adverse effects of some early-life exposures, such as cigarette smoking, on many systems are obvious and beyond debate. But the effects of other exposures may be less obvious and more difficult to define, such as the more-subtle variations in nutritional patterns, microbial exposure, physical activity and individual experiences of stress. Importantly many of these early factors may determine both our *early* programming and our *ongoing* patterns of response and behaviour throughout life and maturity.

WHY HAS THE IMPORTANCE OF EARLY LIFE BEEN OVERLOOKED FOR SO LONG?

The health of both women and children has traditionally not been given the emphasis or the attention that it deserves. One of the great challenges we face is overcoming a traditionally strong, sometimes paternalistic, 'adult' mindset within orthodox medicine, although this is changing.

History plays a big part in this. The field of paediatrics was not really developed as a defined entity until the nineteenth century. That makes child health a relatively new area of medicine. After many centuries of focus on adult anatomy, physiology and disease, understanding of the unique aspects early life was generally lacking.

Throughout the ages, infant and child mortality have always been extremely high. In Europe, even as recently as the sixteenth and eighteenth centuries, the probability of dying within the first year of life was as high as 25 to 40 per cent, and only around 50 per cent survived beyond adolescence; this was not better than the estimated rates for ancient Rome and ancient Greece (circa 500–300 BC).[5]

The dawn of the 'Age of Enlightenment' in the seventeenth and eighteenth centuries saw a greater emphasis on education, as society was reformed through knowledge and science, away from tradition, faith and superstition. Jean-Jacques Rousseau (1712–78) was among those who argued that childhood should be preserved as haven of transient shelter from the harsh realities of adult life. Rousseau was also one of the first to advocate developmentally appropriate education, and divided childhood into different stages. He considered that in the 'first stage', until the age of about twelve years, children are impulsive and guided by their emotions. During the 'second stage', from twelve to sixteen years, they begin to develop reasoning. And in the 'third stage', from around sixteen years, they make the transition into adulthood. This renaissance bought a greater awareness of the sanctity of childhood.

The social, economic and cultural landscape changed radically in the late eighteenth century with the Industrial Revolution. Millions of very young children were forced into the factories and mines, working in appalling conditions to support their families. In Victorian England, child labour continued long into the nineteenth century. Children were far cheaper, and were just as productive as strength and size were less important to operate the 'modern' machinery. With growing pressure from reformists to improve working conditions, the Factories Act of 1833 introduced the first laws against child labour in the textile industry, and by 1878 the Factory Code applied to all trades. No child under the age of ten was to be employed. There was compulsory education for children up to ten years old, and 10- to 14-year-olds could only be employed for half days. This did improve the plight of many children, although child labour continued into the twentieth century in many regions, including other areas of Europe and the United States. Even today, child labour is still prevalent in developing countries, mostly in rural and agricultural settings, where high poverty persists and schooling opportunities remain limited. In 2010, the rates of child labour were highest in sub-Saharan Africa, with several nations having over 50 per cent of 5- to 14-year-old children working.

Concepts of childhood continually evolve. Some argue that the child has developed a more dominant place than ever before. And a more protected one, with much greater collective parental fear of both strangers and the outside environment. Other influences include the changes in family dynamics, working life, a greater role of fathers in child rearing, and new technology and electronic media which change patterns of play.

• • •

It was not until 1802 that the first hospital for children, *Hôpital des Enfants Malades,* was opened in Paris. It was another fifty years before the first English paediatric hospital was founded in Great Ormond Street, London. In the century that followed, paediatrics gradually developed of as a specific field of medicine. And since then we have seen a dramatic and progressive reduction in infant mortality to around 1 to 2 per cent in modern, developed world regions. This occurred in tandem with the industrial revolution, changes in social structures and the gradual development of public-health measures. But the specific focus of the unique issues and vulnerabilities of early life played an essential part. It is truly terrible to still see very high rates of infant mortality in impoverished, developing world regions, where resources and child-health services remain limited. Knowing this is avoidable makes the tragedy ever so much greater.

By the nineteenth century when paediatrics was just emerging, the more 'ancient art' of medicine was already strongly established. Although paediatrics was slowly added into medical training programs, it is still seen as a separate speciality area, and rarely given appropriate emphasis. There is still a dominant focus on adults and on diagnosing and treating established pathology, without a proportionate focus on prevention and early intervention. Medical students continue to spend the vast majority of their time studying adults. Many complete their training without much experience or confidence with children, and generally without strong concepts of the developmental origins of disease. Traditional medical training has generally also been very discipline focused, concentrating on specific organs and systems rather than taking a multisystem approach and looking more at the interconnections, which may come more naturally in paediatrics.

MEDICINE IS SLOW TO CHANGE:

There is no doubt that the medical profession has been very slow to change. An appropriate level of scientific scepticism is, in fact, *essential* for driving the quest for good evidence, so that any change in thinking or practise is carefully justified. We are all encouraged to question and to doubt, and to vigorously test each new hypothesis with the best scientific methods. Another core value is 'masterly inactivity' which reminds us that it is sometimes better to do nothing and to wait for natural resolution of a problem than intervening and potentially making the situation worse. These are wise concepts, which go a long way to explaining the conservative nature of medicine and natural reticence for change.

There is a place for being cautious, but not indiscriminately at the expense of new discovery.

A pertinent example of this comes from the 1840s when the young Hungarian doctor Ignaz Semmelweis showed clear evidence that hand washing with chlorine solutions in the labour wards of Vienna improved the extremely high rates of maternal and infant mortality. Because the idea did not fit the prevailing beliefs of the medical establishment, Dr Semmelweis's call for hand washing was ignored. And tragically, countless women died avoidable deaths before the truth of his hypothesis was finally recognised.

A more modern example comes from my own university (UWA) in the 1980s, when two medical researchers had the audacity to suggest that gastric ulcers and stomach cancer might actually be caused by a microbe called helicobacter.[6, 7] The establishment did not believe that any bacteria could live in the acidic environment of the stomach, despite earlier German studies showing their presence. Taking a rather unorthodox approach to proving his case, Professor Barry Marshall infected himself with the disease-producing helicobacter strain. He became unwell within days, and showed objective evidence of new gastric inflammation on endoscopy. He then treated himself successfully with antibiotics. This was a major breakthrough and soon became standard therapy. Both Barry Marshall and his collaborator Robin Warren were awarded the Nobel Prize in Medicine for this discovery in 2005.

But scientific rigour and caution are in place for a very good reason, and when this fails the effects can be equally disastrous. In 1998, Dr Andrew Wakefield published a report in *The Lancet*, claiming to have identified a new 'gut syndrome' associated with autism and also raised a *possible* link to the MMR (measles, mumps and rubella) vaccine. It was later discovered that the report was fraudulent and that Wakefield was paid by a personal-injury lawyer who planned to seek damages from the vaccine manufacturers for alleged suffering caused by the vaccine. An independent enquiry later showed that the data suggesting that autistic features appeared shortly after the MMR vaccine were falsified. After more than a decade of controversy and investigation, *The Lancet* retracted the paper and Wakefield lost his medical licence. But not before immeasurable damage was done.

In this case the response to new 'findings' was paradoxically rapid. But this was the result of media frenzy and public alarm rather than the medical profession. Infant vaccination was already an emotive area, and Wakefield's comments lit the fuse on this powder keg. As expected, the result was explosive. Vaccination rates fell and public paranoia spread. Hundreds of unprotected children developed complications after infections with mumps, measles or rubella and there were even some deaths. It is frightening that this situation could have occurred in the first place, and that the usual checks and balances failed. Cases like this damage public trust in medical research, to the detriment of all of the legitimate work done by the vast majority of honest, committed medical scientists.

Then we come to another unbelievably bizarre hoax – the 'strange case of Dr Chandra'. Dr Ranjit Chandra was a world-famous medical researcher from the Memorial University of

Newfoundland who had received Canada's highest honour as an Officer of the Order of Canada. He published a series of very impressive high-profile clinical trials including studies of infant formula in allergy prevention, and studies showing vitamin products improve memory in the elderly. These studies attracted much attention and were widely quoted. He had received considerable amounts of money from pharmaceutical and several infant-formula companies to conduct these studies. It was some years before it was suspected that the results were completely fabricated. Ivan Oransky, executive editor of Reuters Health, has been following cases of misconduct, plagiarism and fraud. He has noted, with concern, the significant rise in scientific misconduct, and the considerable delay in correcting the scientific record when fraud is discovered, arguing 'It is not misconduct that kills trust in science. It's the cover up'.[8]

These events have also set the scene for an understandable backlash throughout the profession, with pressure for more accountability, greater scrutiny of data, and full disclosure of all financial relationships and any other conflicts of interest. It means a much higher level of scrutiny and much tighter regulation, and even more conservative standards when considering any new trials, especially for research that involves children.

In some ways this is making a conservative profession even more risk averse. There has also been a changing dynamic between healthcare professions and the community. In the last fifty years the medical profession has actively striven to become less paternalistic, and encouraged patients to take a more active role in the 'therapeutic relationship' with their doctors and allied healthcare workers.

Complementary health providers are also taking a more prominent role in healthcare. There is a much wider appeal for what may be considered to be more holistic. Partly driven by the media, the modern public has been encouraged to question all aspects of 'the establishment' and have become progressively more cynical in the process. We are also much more likely to do our own research through the internet these days.

We must consider this complex modern panorama to understand that medicine and health care are under pressure to change, while at the same time faced with the need for even higher levels of caution and to be more circumspect than ever. So a continued level of inertia is inevitable to some extent. It is a confusing time, but with such dramatic advances in technology and scientific discovery, there is still great optimism and great good will. We are entering an age of unsurpassed knowledge when the need for wisdom has never been greater.

GETTING THE MESSAGE THROUGH:

On too many occasions, when trying to convince my adult counterparts that it would be a good idea to give greater emphasis to child health and early-life origins of disease during medical training I have been met with disinterest. And many of my paediatric colleagues describe similar experiences.

But I am pleased to say that things are changing. A lot. My own speciality, allergy and immunology, spans both adult and paediatric medicine. And I have seen a dramatic shift in emphasis since I started my medical training. In the 1980s and early 1990s research on the early origins of allergy was pretty much on the fringes. Now, the 'developmental origins of allergic disease' plays a prominent role in all international allergy meetings.

This integration of 'early origins' is less evident in other areas of medicine. Every time our medical training curriculum undergoes an overhaul, we paediatricians continue to remind our colleagues working in adult medicine that what happens in pregnancy and childhood is the foundation for *everything* that follows. In the future we hope that all of these things will be considered more in tandem.

In all of this we can see the recurring need to have a larger, longer vision. These philosophies call on us to make investments that might not bear fruit for many years, or even in our lifetime.

We need to be in this for the long haul.

. . .

The chapters that follow focus on a series of examples of specific disease conditions affecting specific organ systems. I hope to reinforce that there are more interconnections than we ever suspected, and that a more holistic view is essential. This becomes even clearer when we discuss how many early-life 'risk factors' affect multiple systems and the risk of multiple diseases.

4

Weight gain and obesity

We are getting fatter. In some regions, more than a third of the population is now obese. The rates of obesity in children and adolescents raise grave fears for the long-term health of this generation. And spiralling rates of obesity in pregnant women amplify the chances of heavier children, further driving obesity and 'overweight' into future generations.

Being overweight or obese is a major risk factor for many chronic diseases, including type 2 diabetes and metabolic abnormalities, liver disease, joint disease, some cancers, heart disease, dementia, stroke and other consequences of vascular disease such as kidney and eye disease. Even early-onset diseases such as allergy are more common in obese children. It is now well recognised that the patterns of weight gain, metabolism and even the total numbers of fat cells in our bodies are determined in early life – making this a very core part of the 'early-origins' story, and a very logical place to start exploring the specific effects of early programming on future health.

Tipping the scales:
The obesity epidemic has built quickly across just two generations. In the USA, rates of obesity have increased from less than 15 per cent in the 1960s to over 35 per cent in 2010.[1] In Australia, more

than 28 per cent of adults are now obese (and 63 per cent are either overweight or obese).[2] There has been an almost universal upward trend since the late 1970s, and the increase appears to be occurring more rapidly in developing regions. In India, rates have doubled in only ten years. And in China the rates have tripled. Global projections by the World Health Organization estimate that 2.3 billion adults will be overweight and 700 million will be obese by 2015.

We officially define 'obesity' and 'overweight' based on the body mass index (BMI), which takes into account our weight in relation to our height: our weight (in kilograms) divided by our height (in metres) squared. There is a normal range of variation, and people with a BMI outside of this range have a greater chance of ill-health. In general, a BMI in the range of 18.5 to 24.9 (kg/m^2) is considered healthy or 'normal'. Less than this is considered 'underweight'. A BMI of twenty-five to thirty is considered 'overweight' and a BMI of more than thirty is considered 'obese'. These definitions have obvious limitations. For example, the BMI cannot distinguish between someone who is very muscular with strong bones and little fat and someone with the exact same BMI who has more fat, thinner bones and poor muscle mass. Clearly one is far healthier than the other, and has a very different risk of obesity-associated NCDs. So BMI must always be considered in the wider context.

BMI measurements also fail to recognise differences in the *distribution* of fat. Fat on the thighs, buttocks, arms and the superficial layers around our body is less concerning than fat around our waistline. Expanding waistlines reflect fat in and around the organs such as the liver and heart, and in blood vessels. Our intestines are insulated, cushioned and supported by fatty tissue (called the 'omentum'). In obese people these deep internal deposits expand massively. This 'visceral' fat adversely affects our metabolism and promotes inflammation in many organs (below).

Another issue is that the definitions that apply to Caucasians may not work as well for predicting disease risk in other races. For example, the risk of diabetes or heart disease for Asians starts to rise with a BMI as low as 22 kg/m^2, which is still in the range that we consider healthy for Caucasians. In this light, the rising rates of obesity and overweight in Asian regions take on even greater significance.

The core reason for making any classification is to identify who most needs interventions to reduce the risk of disease. Someone with a BMI of forty to forty-five is likely to live eight to ten years fewer than someone who is not obese. Obesity increases the risk of almost all NCDs. And we don't have to be morbidly obese to suffer the consequences of excess fat. Preventing obesity and overweight is fundamental to reducing the burden of NCDs.

Again, the earlier the better. Recently, the UK government 'Obesity Taskforce' of the Foresight Group identified early life as the *only time* when effective strategies to combat obesity are likely to succeed.[3]

Eyes of the beholder:

Ironically, as our populations all get fatter, popular culture in Western societies is beset by an overwhelming obsession for thinness, even though this is rarely sustainable – and is certainly not ideal, normal or necessarily healthy.

We must remind ourselves that this concept of 'thin' beauty is a modern social construct. Our often futile attempt to aspire to an 'ideal' of thinness promotes unhealthy eating patterns, body image and behaviour, which in turn contribute to eating disorders such as anorexia and bulimia, particularly in girls and young women.[4]

On the whole, the messages we get about our bodies through media and advertising are intended to appeal to our human desire to become 'more beautiful' rather than 'more healthy'. Much of the media hype reinforces the idea that it is 'our fault' as individuals if we are fat, which is not necessarily true, as we will see later. We need to change this message and understand the deeper complexities of this issue, for the sake of our future health and sanity.

This book is intended to provide a broader, deeper and more realistic view of this weighty challenge, so that we are better equipped to deal with it. Specific exercise routines and dietary changes have been explored by others, such as Timothy Caulfield in his book *The Cure for Everything: Untangling Twisted Messages about Health, Fitness and Happiness;*[5] for those looking for a 'how-to' plan of what works and what doesn't, I recommend it. Here though, we will explore more of the *origins* of this problem, *how* 'fat' is damaging our health and *why* this is harder for some of us to lose weight than others.

LIVING IN CAPTIVITY:

The fact that obesity rates have increased so significantly in the last thirty to forty years points to large-scale societal changes in our behaviour, diet and lifestyle. These environmental changes are clearly driving the complex metabolic changes associated with obesity, and underscore the fact that we are not just getting bigger, but that there has also been a gradual shift in our underlying metabolism and physiology. In fact, we show all the hallmarks of animals living in captivity.

Michael L. Power and Suzette Tardif[6,7] studied a colony of marmosets living in San Antonio Zoo and observed that over several generations, despite no clear changes in animal management, the animals got progressively fatter. These primates showed all of the metabolic changes seen in human obesity and provide a very useful comparison. In particular, the researchers saw that the development of obesity occurred at a very young age and was associated with even early differences in body-fat composition and growth patterns. Maternal obesity was an independent risk factor for obesity (over and above the postnatal diet), suggesting that some of the programming for obesity may occur even before birth.

There are interesting parallels with humans. We have built ourselves a captive environment, with a steady food supply and stable temperatures, free of predators and with little need for physical exertion.[8] Sugar, once scarce, is now abundant. Our bodies are programmed to store this energy source as fat to help our future reserves. And we are becoming increasingly efficient at that, to our detriment.

But Power's earlier work with golden lion tamarins offers a glimmer of hope. This study showed that after releasing the animals back to the wild it took several generations for them to return to their thinner, 'wild' appearance. But it eventually happened. This demonstrates adaptability, and reversible plasticity, although the reversal took several generations. But will humans want to go back to 'the wild'? We have grown rather fond of our cages. I rather like mine. Somehow, we need to find a way to enjoy our modern perks while living more closely with nature as though we were 'free'.

WHAT ELSE IS DRIVING OBESITY IN MODERN ENVIRONMENTS?

The most obvious drivers of obesity in captivity are too much food and too little exercise. But other factors exacerbate the risk of weight gain. Some have significant effects before birth and early childhood. Some maternal exposures, such smoking in pregnancy, which reduce birth weight, actually increase the risk of obesity in later life. Other factors are unique to early life, such as exposure to gestational diabetes and the metabolic effects of maternal obesity on foetal programming.

Childhood is filled with obesity risk. Many children no longer walk to school and the use of electronic media has reduced activity across all ages. Increased 'time spent sitting' is a risk factor for later ill health (Chapter 8). Artificial light from electronic devices can interfere with the natural production of melatonin by our brains, increasing sleep disturbance. Recent studies in 10- to 12-year-old school children have shown that reduced sleep is a risk for higher BMI and increased risk of obesity.[9] Stressful environments can also result in sleep disturbance, psychological problems and altered eating patterns. The significant increase in children developing mental, psychological and behavioural disorders (Chapter 7) is also likely to be contributing to the risk of obesity; and some of the medications used to treat psychological disorders promote weight gain. Other social changes, such as pregnancy in older mothers, may also increase susceptibility to obesity. Many of these factors are interrelated and it is difficult to tease out prime causes, which will vary between individuals.[10]

There are other, less-obvious factors that are more difficult to prove. These include environmental pollutants including 'endocrine disruptors' (Chapter 9) that can interfere with fat metabolism, and more stable temperatures in our housing which restricts the need to burn fat for warmth. Cleaner, more hygienic environments and changing dietary patterns may have significant effects on the patterns of our gut bacteria – with surprising effects on our both our fat metabolism and immune function, as discussed below.

Because there are so many different, often unique, factors at play (including intergenerational patterns) it has become increasingly difficult to untangle the biological changes associated with modern lifestyle, making causes and effects quite difficult to predict.

MORE THAN JUST A MATTER OF SIZE:

Fat is much more than just a simple storage system. It is a highly active part of our metabolic system which can influence the metabolism and function of many other body systems. It contains immune cells capable of driving inflammation. It produces hormones that can influence our behaviour, mood and appetite. And, if in excess, fat can be a source of toxic 'free radicals' which can damage other organ systems, including the developing foetus. This may be one of the reasons that a high caloric burden over a lifetime also appears to accelerate ageing (Chapter 11).

It is not surprising that obesity is associated with so many chronic NCDs as a result of inflammation, changes in metabolism and organ damage – which lead to what is referred to as the 'metabolic syndrome'. Having some fat is normal, and is an essential part of our metabolic processes. Every gram of fat we 'burn' gives us roughly nine calories of energy, and fat as a store of energy allowed our ancestors to survive with an unpredictable food supply. Because fat is the most efficient form of energy storage, when we take in excess calories from carbohydrates, protein and alcohol these are logically converted to fat.

There are also different kinds of fat. Most of the fatty deposits that cause disease are the more common 'white' fat. The other kind of fat is 'brown' fat, which is more common in newborns – it generates heat, protecting babies from hypothermia. Because it is designed for *burning* energy for heat rather than *storing* energy it is considered 'healthier' fat. The proportion of brown fat declines with age, but can be increased in adults in some situations, such as exposure to colder temperatures.

At the most *basic* level, we can see that obesity develops when there is a persistently positive energy balance: more energy intake than energy expenditure, leading to fat storage. But there is *far more* to it than that. Many of the factors that lead to obesity have complex metabolic, physiological and immune effects that accentuate the 'metabolic syndrome' of obesity, and actually make obese people *even more efficient* at storing energy as fat than non-obese people.

PHYSICAL ACTIVITY – MORE THAN JUST FAT BURNING:
The declining physical activity associated with Western lifestyles is implicated in accelerated ageing, increased organ failure with stress, impaired immune and vascular function, and heightened brain ageing. It is, therefore, a major risk factor for a broad range of NCDs.[11]

Physical inactivity is strongly associated with chronic low-grade inflammation in otherwise healthy people. 'Sitting time' is also a specific risk factor for many NCDs. On the other hand, regular exercise protects against 'all-cause' mortality, particularly through prevention of NCDs.[12] Physical exercise increases the release of growth hormones and other hormones that promote the health of the immune and nervous systems and improve our responses to stress. It has general benefits for mental health and reduces the risk of depression and anxiety.

Physical activity also has direct anti-inflammatory effects, counteracting or reversing many of the inflammatory 'cytokines' seen in obesity.[13] These and other physiological changes also help protect against cardiovascular disease and type 2 diabetes mellitus.[14] During exercise, muscle contraction induces release of other newly recognised immune mediators (now called 'myokines'), which enhance the breakdown and metabolism of fat.[15]

In the context of early programming, physical inactivity in pregnancy and early childhood may be yet another factor contributing to the risk of low-grade inflammation during early development, with long-term consequences on health.

MICROBIAL INDUCED OBESITY – UNEXPECTED EFFECTS OF UNHEALTHY DIETS:
The food we eat is also food for the bacteria in our guts. Eating less dietary fibre can alter the proportions of 'friendly' bacteria in our guts, and potentially increase the bacterial species that promote

inflammation. Gut bacteria are also important for 'fermenting' dietary fibre to release anti-inflammatory short-chain fatty acids which are very important for intestinal health. If we eat less fibre, we have less of these protective fermentation products and the intestines become leakier at a microscopic level, so that bacterial products ('endotoxins') from more pro-inflammatory bacteria more easily seep into the blood stream. This produces higher levels of inflammatory immune products (called 'cytokines') in the blood stream of obese people. There are also metabolic changes in the liver, switching on the machinery that more efficiently converts our dietary fuel into fats ('triglycerides'). These can be transported through our blood stream to fill the many fat cells ('adipocytes') around our bodies. At the same time, there are changes in our glucose (blood-sugar) metabolism. We become more resistant to the effects of insulin, which normally regulates our blood sugar. The result is that our blood-sugar levels rise, but our cells are still hungry and still need fuel (Chapter 5). Other hormonal changes drive food craving so that we want to eat even more. In essence, these metabolic changes make us into highly efficient fat-making and -storage machines.

Clearly, if these metabolic derangements occur during pregnancy there can be significant consequences for foetal metabolism and the future risk of obesity. Both maternal obesity and maternal diabetes are now common in pregnancy, and compound the risk to the next generation (below).

The interesting thing is that obese people and animals have been shown to have different patterns of bacterial colonisation ('microbiota') in their intestines. In one experiment, when the microbiota from the intestines of obese mice was introduced into non-obese mice they became obese, but this did not happen if they were given bacteria from lean mice.[16] We are only just beginning to understand how the billions of 'foreign' bacterial genes (called the 'microbiome') contained in our guts have a pivotal influence on our immune and metabolic health. But diet is still the key.

In humans, there is good evidence that microbiome patterns are established quite early in development, and are influenced by maternal colonisation, breastfeeding, early infant dietary patterns and the background environmental context. After this, the microbiome pattern is relatively stable for the rest of life, but may vary with gut infections and major dietary changes.

But before we focus in on the early-life aspects of this story, it is useful to understand more of the metabolic process that led to weight gain and obesity, as many of these also have implications during foetal life and early childhood.

REGULATING OUR FAT:

Most systems in our body are tightly regulated around an optimal set point or 'normal range'. A balancing process referred to as 'homeostasis' applies to the mechanisms that control levels of proteins, hormones and solutes in our blood, as well as other processes such the 'thermostat' that controls our temperature. Our fat or 'adipose' tissues have physiological systems to achieve homeostasis and there are a number of counterbalancing hormones that work to regulate our levels of fatness or 'adiposity'. Some scientists refer to this as our 'adipostat'. This involves multiple mechanisms and hundreds of genes that control appetite, metabolism, behaviour and weight gain.

Fat produces its own hormones. These are important for controlling or suppressing further fat deposition. The best known of these is 'leptin', which is released into the blood stream and circulates to influence other organs. In the brain, leptin suppresses appetite. In the pancreas, leptin regulates insulin production to control our blood sugar. It can also protect from high blood pressure and vascular disease. In other words, it works to counteract most, but not all, of the features of the 'metabolic syndrome' and protect us against obesity.

Children born with a leptin defect are compelled to eat, driven by an unlimited appetite. Synthetic leptin has helped save these children. When artificial leptin was first produced there were hopes that it would be the miracle cure for obesity. However, this was followed by disappointment.[17] This is because most people with garden-variety obesity actually already have *higher* leptin levels but have become resistant or tolerant to its effects.

'Ghrelin' is one of the main hormones that counterbalances the effects of leptin. It is produced in the stomach when it is empty and circulates to our brain, where it stimulates the 'feeding centres' (in our hypothalamus) that induce hunger. It also inhibits appetite suppression by leptin in our fat cells. Ghrelin also has effects on the immune system and blood-sugar control, and leptin and ghrelin appear to reciprocally regulate each other's activity in many other organ systems.

Because these hormones are designed to maintain the 'status quo', they can seem to work against us. When our fat stores start to reduce, the levels of ghrelin start to rise. This makes us feel very hungry. As leptin levels subsequently fall, our energy expenditure, nervous-system activation and thyroid hormones decrease (slowing metabolism) to collectively drive us to regain weight. These hormonal changes persist *sometimes for years* after we lose weight.

If we have been overweight for a long time, or had more fat cells at birth or in childhood, our 'normal' set point will be high. Our BMI will be too high, but our metabolism thinks that is 'normal'. And every time we try to lose weight, the hunger hormones kick in to save us from starvation. So now we can see how important our early programming can be.

But it gets worse. As we get fatter our leptin levels actually start to increase, but they have less effect. Our appetite control works less well. As the 'metabolic syndrome' sets in, we also become resistant to insulin, which is the 'key' that lets glucose in the 'door' of each cell (Chapter 5). Glucose is an important fuel, especially for our brains. We are fat, but our cells are actually starving. Our blood glucose climbs and diabetes sets in. As we become more obese and the metabolic syndrome is more developed we start to see abnormal liver function, heart disease and the many consequences of chronically high cholesterol and high blood sugar.

It remains logical to try and restore the balance with a long-term program of diet and exercise. But we need to understand how hard this can be for people, and be realistic about what can be achieved and how long it may take. We are working against a biological program that makes weight loss much harder, especially when that biological program is well established. That is why it is important to intervene very early in the process or prevent it in the first place. It is also important to understand that many of the factors that lead to obesity begin before we have any conscious control. And to stop blaming individuals for driving the modern health crisis.

There are numerous examples in animals where calorie restriction, without malnutrition, improves many aspects of health and increases longevity (Chapter 11).[18, 19] These effects appear to be mediated, at least in part, by insulin-like growth factor 1 (IGF-1), which plays an important role in metabolism, growth and development. Clearly this is about moderation and common sense. Extreme calorie restriction is *not* the solution to our current health crisis, and can be very harmful. In particular, restriction during pregnancy and early life is actually a significant *risk factor* for NCDs and obesity in the offspring, as explained later on. The emphasis should be on a healthy, balanced diet.

Obesity and vitamin D – is there a link?

Low vitamin D levels in pregnancy and infancy cause rickets, increased susceptibility to infections, and many other health problems. While the more profound effects of this deficiency, such as rickets, are now rare in industrialised nations, vitamin D deficiency has recently re-emerged as a growing issue in many developed countries. This is thought to be due to more time spent indoors, use of sunscreens and hats, and covering skin. This deficiency has been associated with metabolic and immune effects, as well as a wide range of NCDs including rising rates of allergy, heart disease, some cancers, multiple sclerosis, rheumatoid disease, and other autoimmune diseases. At particular risk are darker skinned people, women who cover skin for cultural reasons, and the obese or overweight.

There are a number of possible reasons for the lower vitamin D levels seen in obesity.[20] It could be because vitamin D is fat soluble, and attracted or 'sequestered' into the fat stores from the circulation. It could also be due to less gut absorption, or less outdoor exercise and time in the sun where vitamin D is activated in our skin. It could simply be because vitamin D is 'diluted' in a larger body volume. Whatever the cause, the association is well established, and higher doses of vitamin D are required to correct low vitamin D in obese people.

A recent study in obese adolescents showed that correcting vitamin D insufficiency with supplementation had some metabolic benefits, even though there was not overall difference in weight (BMI). After six months, the adolescents who received the vitamin D supplementation showed improved insulin sensitivity and favourable changes in the hormones (the ratio of leptin to adiponectin) that regulate fat metabolism.[21] This suggests that correction of poor vitamin D status (if present) is an important addition to the standard treatment of obesity.

The role of vitamin D supplementation in pregnancy is at the forefront of many discussions around how to provide optimal conditions for early programming and disease prevention. At the present time this is a very murky area (and is discussed more in Chapter 8).

GENES VERSUS ENVIRONMENT – WHAT IS MISSING FROM THE EQUATION?

Obesity, like all complex disorders, is often discussed in terms of 'gene × environment' interactions. But one very important factor is missing, and that is 'timing'. To really advance the understanding of modern diseases it is critically important that we add timing, or 'development', to this equation.

It is immediately clear that there are individual variations in the 'adipostat' which determines our constitutional sensitivity to excessive weight gain. But what causes these individual variations? Traditionally this would have been attributed entirely to 'genetics'. But even when we take genetically *identical* animals and put them in *identical* 'obesogenic' environments, we can still see differences in their weight-gain patterns if they experienced subtly different events during their early development (as discussed in more detail below). In humans there is also emerging evidence that developmental programming can modify the programming of our 'adipostat' and risk of obesity and other NCDs.

So, just how much of this is genetic? How much is our early environment? Is it even possible to separate the contribution of these factors?

There are more than 100 different genes that are involved in regulating our weight, each with a relatively small role to play. Some of us carry slightly different versions of some genes, and this makes us more likely to develop obesity than others (see 'genetic polymorphisms' in Chapter 11). For example, some versions of the 'FTO' gene (fat-mass and obesity-associated gene) give us a stronger predisposition to weight gain. People who have two copies of the 'risky' version of the gene weigh on average 3 kg to 4 kg more and have a 1.67 times greater risk of obesity than those without it.[22] And it is likely *many* genes operate in a similar way. These genetic differences may not cause a problem, however, *unless* we are in an environment that brings out their worst.

Even with 'risky genes' we are unlikely to develop obesity living a traditional lifestyle in a traditional environment. We need to be in an 'obesogenic environment'. *Only* our changing environment and lifestyle can explain the recent dramatic rise in obesity over the space of forty to fifty years. Our genetic code has evolved over millennia, and everything we know tells us that our genes could not have changed in just the last forty to fifty years – but have they?

We have recently discovered that our genetic program *can* change much more dramatically and quickly than ever imagined. And that this can *easily* occur within a few generations. Although the sequence of our genes does not change in that timeframe, the patterns of genes that are expressed can change dramatically.

Our genes are just blueprints that we use to create all the many proteins that make up our bodies. Because each cell type needs to create different proteins to make up the different kinds of tissues in our bodies, the program that controls this is all-important. The 'epigenetic program' tells each cell *which* genes to express, as well as *where* and *when* to express them (see 'epigenetics' in Chapter 11). Our genetic sequence determines which proteins each cell *can* make, but our epigenetic code determines what we *actually* make. When scientists embarked on the Human Genome Project to discover and sequence the full human gene code there was great anticipation that this would reveal all the mysteries of human health and disease. But, of course, cracking the genetic code only revealed a part of the picture, and many more questions. In a sense it was like discovering a vast database of numbers, but without the program to read it. The discovery of the epigenetic code has been another major leap forward in understanding how our genes work, but there is still a great mystery to solve.

The discovery of the epigenetic code revealed how the primitive stem cells of a newly formed embryo know to develop into all the different tissues that make up our bodies. It also revealed how our gene expression can *change* in response to our environment. The environmental conditions of the mother are 'sensed' by the embryo; this can lead to a change in its pattern of gene expression, which can permanently alter shape, form, physiology and behaviour in anticipation of particular environmental conditions. There are some quite dramatic examples of this in animal studies. In humans, there is now some evidence that early environmental changes have epigenetic effects that result in physical changes, altering our future health and disease risk.

So how does any of this relate to the risk of obesity? A number of mice studies have shown that modifying the environment *very early* in life when mice are still developing can radically alter their colour, appearance, eating behaviour and weight as adults.[23-25]

In one set of experiments, scientists exposed pregnant agouti mice to a chemical found in plastic bottles, linings of canned food, and dental sealants called bisphenol A (BPA). There were radical effects on the offspring. Exposure to BPA in pregnancy induced epigenetic changes, switching *on* the 'agouti gene' in her offspring. In this mouse the agouti gene regulates coat colour, as well as feeding behaviour and the body weight set point, predisposing to obesity, diabetes and cancer. Animals exposed to BPA in pregnancy developed yellow fur and were more prone to these conditions compared to the small, brown offspring not exposed to BPA. In other words, they look *completely* different even though their gene sequence is *identical* to non-exposed mice. BPA exposure in foetal life reduced 'methylation levels' of a number of genes. Methylation is a way that the epigenetic program keeps genes in an 'off' position so they are not expressed (Chapter 11). Basically, by reducing the level of methylation, BPA switched 'on' genes that are typically 'off'. The researchers also found that other environmental factors could protect against the effects of BPA by counteracting the reduction in DNA methylation. These protective factors included folic acid, vitamin B12 and the soy phytoestrogen, genistein.[26] Importantly, many of the consequences of altered 'foetal programming' were not apparent at birth but had long-term effects on disease risk for the mice once they became adults.

These epigenetic mechanisms provide a critical link, showing *how* the early environment can influence adult obesity and associated NCDs. It also raises questions about the possible effects of BPA on modern humans who eat plastic- and tin-packaged food and beverages. These latter questions are harder to answer in humans than they are in mouse models where all other variables can be controlled. But US studies show 90 per cent of women of reproductive age have BPA detectable in their urine. And BPA levels in pregnancy have also been associated with metabolic changes in human newborns (more in male babies)[27] and behavioural effects (more in girls),[28] although the long-term effects are not clear. Notably, in both the USA[29] and China[30] BPA levels in children have been associated with obesity. It is logical to try to reduce unnecessary exposure to as many chemicals as possible, but it will always be difficult to prove cause and effect in humans, because there are so many other factors in play.

We can now be fairly sure that many dietary changes, many modern pollutants and even the changing microbial patterns are having effects on our early patterns of gene expression through epigenetic effects. As there are *so many* modern environmental factors, we should not become preoccupied with just one or two, and make sure we look at the *whole picture*. I selected the example of BPA here because it is well illustrated by both animal experiments and human observations, and because it relates to early programming for obesity. There are certainly a whole host of other factors that play a role in the escalating tendency for human obesity. And we have a lot of ground to cover in discovering how these all interact, and to what extent they are altering our epigenetic program in early life.

MATERNAL UNDER-NUTRITION CAN PROGRAM FOR FUTURE OBESITY:

Stunting and growth restriction due to poor nutrition have been long been major issues in the poorest regions of the world. Ironically, the small babies of undernourished mothers are at *higher* risk of obesity and other NCDs later in life. Ravelli and colleagues studied the long-term effects of the Dutch 'Hunger Winter' of World War II (described below), and were among the first to describe how maternal starvation substantially increased the risk of offspring becoming obese in later life.[31] The foetal 'epigenetic machinery' is designed to sense maternal conditions, including the mother's nutrient supply, and respond accordingly. If it detects 'scarcity' it will switch on the genes for efficiently storing energy. The foetus is preparing itself for an environment of 'scarcity' when it is born.

If there is significant mismatch between what we have *antici-pated* based on our in-utero experience and what *actually* happens after birth, our risk of disease in later life is much greater. For example, if we encounter an *over*abundance of food after birth when our metabolism has been programmed for scarcity, we are geared to store fat and conserve every calorie; much more than if we had not experienced scarcity as a foetus.

This has been very elegantly shown in animal studies by Mark Vickers and his colleagues in New Zealand. They observed that baby rats born to *under*-nourished mothers ate much more and were more likely to develop high blood pressure and metabolic syndrome than those born to well-nourished mothers, even if they had the same standard calorie diets after birth.[32] When the baby rats were given high-calorie diets after birth, the effect was even more dramatic. These animals had striking differences in their levels of the leptin hormone that regulates body weight. This study showed two important things. Firstly, it showed how patterns of maternal nutrition in pregnancy can 'program' later differences in the feeding behaviour and weight gain in the offspring. And secondly, it provided experimental evidence that 'mismatch' between foetal nutrition and postnatal nutrition can amplify the risk of obesity.

A number of other maternal factors may subtly influence birth weight and, by extension, may also be risk factors for obesity and NCDs. Both older and teenage mothers have smaller babies. Twins and first-borns tend to be smaller. And shorter mothers with smaller pelvic dimensions also have smaller babies. In each of these situations, the foetus is 'sensing' the maternal conditions and responding accordingly. If the foetus grew with no regard for the size of its mother's pelvis neither would survive the birthing process. We see this with donor embryos in IVF. Once implanted, the growth of the embryo follows more closely the conditions of the recipient mother than the donor mother, and birth weight is more related to the recipient's weight than the donor weight or the birth weight of the donor's own children.[33]

Various factors influence the degree to which foetal growth is 'restrained' in pregnancy. The greater the restraint the greater the risk, particularly when babies enter an abundant environment after birth. So even in the more 'developed' regions of the world, high proportions of babies are born after experiencing 'restraint'; as families get smaller, as mothers have their first babies later, as the proportion of babies conceived with ART (IVF) increases (with higher rates of twins), and as rates of prematurity increase. And in all of these situations there is some evidence of an increased risk of future NCDs. However, none of this is absolute! And there are many variables that will continue to influence individual susceptibility over the course of their life.

Some of the clearest evidence to support the impact of foetal growth 'restraint' comes from studies of the long-term effects of a terrible famine known as the 'Dutch Hunger Winter' that occurred towards the end of World War II. A German blockade, coupled with an unusually early and harsh winter, meant extreme starvation for the most populated western regions of the Netherlands.

As expected, children of the women who were pregnant during this horrific famine were smaller, but they were also more susceptible to obesity, diabetes, cardiovascular disease, kidney disease, schizophrenia and many other health problems much later in life.[34] Because the Dutch famine occurred at very specific, short period of time it has been possible to compare the effects of famine at different stages of pregnancy. The children of women who experienced the famine *early* in pregnancy were more likely to develop higher cholesterol levels,[35] high blood sugar and diabetes,[36] obesity,[37] heart disease,[38] schizophrenia,[39] stress responses and some cancers as adults. Interestingly, some of these metabolic effects (on blood lipids and blood sugar) occurred independently of the size at birth or adult obesity. In other words, it showed that foetal adaptations that allow us to *preserve* our growth during maternal malnutrition can *still* permanently affect our adult health with adverse consequences. There were also consequences for people whose mothers experience the famine in *mid* pregnancy and *later* pregnancy, including chronic lung diseases, kidney disease and diabetes.[40]

Perhaps the most startling discovery was that sixty years later there were still detectable differences in the epigenetic program (detected as methylation of the DNA) of people who had been exposed to famine during the Dutch Hunger Winter around their conception and early in their foetal life.[41] Researchers focused in particular on the insulin-like growth factor II (IGF-2), and found that the gene was significantly more 'switched on' (less methylated) six decades later compared to their siblings who did not experience the famine.[42] The changes were not seen in people who were only exposed late in gestation, again showing that timing is critical. But there is much more to understand. That study focused on one particular gene. We still need to examine the much wider epigenetic effects on many other genes involving many other systems, and to understand how other environmental factors can alter our early epigenetic programming. In particular, we need to understand the effects of more modern exposures, including over-nutrition.

AT THE OTHER EXTREME – MATERNAL OBESITY ALSO PROGRAMS OBESITY IN THE NEXT GENERATION:

A woman born thirty years ago in India, growth restricted because her own mother suffered under-nutrition, has a higher likelihood of becoming obese in a more prosperous environment than if she was not growth restricted as a foetus. She also has a much higher risk of developing gestational diabetes when she becomes pregnant herself. Her tendency for excess weight gain and higher blood sugars during pregnancy will increase the risk of her own children being overweight or obese by the time they reach adolescence.[43] And so, the cycle of obesity and increased disease risk perpetuates into the next generations.

The precise mechanisms of these effects are still not fully understood, but many are interrelated. Overweight women are more likely to develop gestational diabetes. This metabolically altered foetal environment with higher blood sugars leads to bigger babies. And high birth weight itself is associated with later obesity. But it is not clear whether it is the maternal diabetes or other factors which drive the development of obesity later in childhood in this setting.[44]

Studies suggest the development of obesity is not just genetic. Researchers from Brooklyn studied 172 children of women who had undergone bariatric surgery (gastric banding) to achieve substantial weight loss. They compared children that were born *before* their mothers had surgery (with an average BMI of 48 kg/m^2), to children of the *same* mothers who were conceived *after* the women had lost weight (to on average 31 kg/m^2).[45] The rate of obesity was dramatically lower in children born after their mothers lost weight, compared to their older siblings. This indicates that the risk of obesity *is* modifiable during early development, and not solely determined by the mother's genes.

CONTROVERSIES OVER INFANT FEEDING GUIDELINES:
At the time of writing this, there is very little consensus and still much debate about the best time to start solid foods.

There are a number of issues at play and quite a bit of history to this story. In the 1970s and 1980s use of infant formulas in under-privileged countries contributed to infant death and disease from malnutrition and infection when the water used to make it was contaminated. And to make matters worse, when a mother stops breastfeeding she is also more likely to conceive again, diverting her already limited resources to her next child. For these reasons the World Health Organization recommends exclusive breastfeeding for at least six months. This recommendation restricts all non-breastmilk intake, even water, but allows rehydration solutions, medicines, vitamins, minerals and other treatments if needed.

While these global recommendations have been very much targeting infant mortality in the developing world, the obesity epidemic has more recently entered this agenda. There are concerns that early introduction of carbohydrate-dense weaning foods could be promoting obesity in more affluent regions[46] through effects on both metabolic programming and taste preferences. There is some evidence that flavours from the mother's diet during pregnancy are transmitted to amniotic fluid and swallowed by the foetus, and that the same flavours can also later be experienced by infants in breast milk. Again, this has not been proven in humans, but animal studies suggest that increasing caloric intake in the immediate postnatal period increases the risk of higher body weight and overeating later in life, and that higher carbohydrate intake is a particular risk factor.[47]

So, in more affluent nations, preventing both obesity and allergies have been used as other reasons not to introduce complementary foods before six months, although recent evidence suggests that delayed complementary feeding until after six months may actually increase rather than decrease the risk of allergic disease (as discussed further in Chapter 10). All this means that parents are faced with conflicting advice.

The specific role of the duration of breastfeeding and the timing of starting solid foods in the risk of obesity is also unclear. One major study to address this followed 10,912 individuals, from low- to middle-income countries, over several decades.[48] They found no differences in the rates of obesity, overweight or diabetes between adults who were 'ever' breastfed compared with those who were never breastfed. There was also no relationship with the duration of breastfeeding. Earlier introduction of complementary foods was associated with higher adult adiposity, but this effect was quite small. Almost all of the studies to examine the long-term effects of feeding choices are 'observational', which means that they have inherent biases and can be difficult to interpret. One of the only situations in which babies have been *randomly* assigned to receive either breast milk (from a human milk bank) or a formula was a study on preterm infants.[49] When these children were followed to adolescence there were no differences in body weight and obesity, but some markers of cardiovascular risk (blood lipid and inflammation profiles) were lower, suggesting some beneficial relationships. For ethical reasons it is very difficult to do the same breastfeeding studies in healthy term infants. One recent study in a high-income country did randomly assign full-term infants to receive complementary food in addition to breast milk to infants' diets from 4 months of age, without showing any effects on growth rates compared to exclusively breastfed infants.[50]

While there is little debate that breastfeeding should be promoted for its many benefits, we cannot claim with certainty that it *specifically* prevents either obesity or allergy. The optimal time to start complementary feeding also needs more investigation. There are several randomised clinical trials of 'early' versus 'later' complementary feeding currently underway that are designed to assess the effects on allergic disease; the effects on weight gain will also be of great interest.

WHERE TO FROM HERE?

As we come to understand that the trajectory for weight gain is set in early life, we can be more realistic about the great difficulties we face in diverting it. None of us evolved to live in 'captivity' and we all face a higher risk of NCDs in this modern environment. Our current crisis shows how a changing environment can change our destiny. But we need to understand that changing these patterns will take more than a generation. And, if we don't address the early-life origins of obesity then it will take even longer to overcome this.

Ending the 'blame game' has to be a major goal. We can take *some* responsibility for our choices and behaviour, but we do not have any control over our genes, our wider environment, or the many of the early-life factors that have programmed our risk of obesity. And it is equally wrong to blame mothers. Every mother is part of a complex social, cultural and economic system that determines her education, her opportunities, her choices and her access to healthcare.

This is a cycle that can only be broken by taking a 'whole of life' approach to prevention, aiming at every age and every stage. And this must all be done hand in hand with improving the education, rights, and health of women.

5

Diabetes and metabolic liver disease

It is hard to believe that a disease could sweep the globe killing millions of people, almost unnoticed. Diabetes is a slow and 'silent' killer and, because its effects are initially subtle, its seriousness is often not appreciated – until it is too late. It is another consequence of modern lifestyle change, and another critical part of the NCD story.

In high-income countries such as the United States, 40 per cent of the population over sixty-five has diabetes or the metabolic changes of 'pre-diabetes'.[1] The World Health Organization estimates that more than 347 million people currently have diabetes, and that 80 per cent of diabetes-related deaths occur in low- or middle-income countries.[2] Type 2 diabetes, which accounts for around 90 per cent of all forms of diabetes, is strongly linked with obesity. Like obesity, the rise in diabetes is associated with Western lifestyle, and there has been a staggering increase in Asia, Africa and South America with progressive economic development. A 2012 study suggested that 334 million Chinese adults and as many as 27.7 million Chinese children are likely affected by pre-diabetes or diabetes,[3] and that 42 per cent of Chinese children have problems such as high blood sugars, cholesterol or blood pressure.[4] India is not far behind, with the second-highest rate of diabetes in the world. Other races are also particularly vulnerable, including Indigenous Australians[5] and Pacific Islanders.[6]

As with other NCDs, early-life factors are important in the predisposition to all forms of diabetes. The surge in gestational diabetes (developed in pregnancy) has increased the risk of future diabetes for mother and child. And rates of pre-existing diabetes in pregnant women doubled in the USA between 1999 and 2005,[7] reflecting worsening health trends, with long-term consequences, again for both the mother and her child.

WHAT IS DIABETES?

Diabetes 'mellitus' is associated with higher-than-normal blood-sugar (glucose) levels. The most common kinds of diabetes mellitus include gestational diabetes, juvenile-onset (type 1) diabetes, and previously-named adult-onset (type 2) diabetes.

High sugar levels spill over into the urine causing the symptoms of frequent urination ('polyuria'), thirst ('polydipsia') and hunger ('polyphagia'). In the longer term, if not treated, high sugar levels cause damage to many organ systems and lead to premature death. The term 'diabetes' first appeared in ancient Greek texts around 230 BCE and was used to describe the excessive passage of urine. The term 'mellitus' (from honey) was added later to describe the sweetness of the urine (glycosuria).

Fundamentally, this is caused by failure of the insulin system to regulate glucose levels. In some cases, this is due to insufficient insulin production (type 1 diabetes), and in others cases body tissues become resistant to its effects (type 2 and gestational diabetes). Insulin is a hormone produced by the islet cells of the pancreas. Its main function is to promote the uptake of glucose into most of the cells of our body. If this system is not working properly, glucose builds up in the blood stream while the cells remain starved of their fuel supply.

Insulin is normally released after eating when an increase in blood glucose is detected. As well as facilitating glucose uptake into cells, higher insulin is the major signal for anabolic metabolic processes which essentially 'build up' fat storage, synthesise proteins and are responsible for cell growth and replication. When fuel is abundant, insulin also promotes conversion of glucose into glycogen in liver and muscle cells so that it can be stored as a backup supply of glucose if dietary levels fall.

When insulin levels are *low*, metabolic processes shift towards a more *catabolic* state of 'breaking down' supplies to release energy for fuel. When glucose (and insulin) levels fall, glucose that has been stored as glycogen in the liver is released into the bloodstream (driven by another hormone called glucagon). Conditions that drive down insulin – such as fasting, low-carb diets, or intense exercise – trigger the release of fatty acids and ketones as alternative fuels. If this state continues, a range of other breakdown products (such as amino acids) can be used to generate glucose in the liver (a process called 'gluconeogenesis').

So, in diabetes when insulin is too low (type 1 diabetes), or cells do not respond to it (type 2 diabetes), we see many of these effects. Glucose will not be absorbed sufficiently by cells, or stored in muscles or the liver. We see many of the metabolic derangements associated with a shift to a catabolic state, with muscle breakdown and the release of fatty acids and ketones. Over time, high sugar levels in the blood and in the tissues lead to kidney failure, blindness, heart disease, stroke, cognitive decline, thrush and other infections, muscle wasting, and poor circulation with associated nerve damage in the extremities. Amputations are common in older people with poorly treated diabetes.

Each of the major forms of diabetes is increasing. They are linked to modern environmental changes (diabetes is rare in traditional environments), and they are each linked to inflammation and metabolic dysregulation. But the underlying pathological processes for the 'types' are different.

Type 1 diabetes is an immune disease:

Type 1 diabetes is still the earliest form of diabetes to appear, and a logical place to start examining the 'early-life origins' of diabetes. It is caused when the islet cells in the pancreas fail to make enough insulin. The most common cause for this is the immune system generating antibodies against the islet cells. This form of immune-mediated disease is known as 'autoimmunity' because it is self (auto) destructive. Antibodies are designed to target very specific proteins such as the elements of viruses and bacteria, and it is not clear what causes the immune system to mount a misdirected attack against 'self' tissues.

The auto-antibodies in diabetes are specific for the islet cells and leave all other tissues unscathed. While genetic predisposition is a factor, modern environmental changes affecting the immune system appear to be contributing to a recent rise in risk for this disease.

Unlike those who develop type 2 diabetes, people with type 1 diabetes are not usually overweight at the onset of their disease. Treatment requires replacement of insulin by injection or infusion, and people usually respond well because their sensitivity and responsiveness to insulin is generally not affected, especially in the early stages of the disease. That is why this form is also referred to as 'insulin-dependent diabetes mellitus' (IDDM).

Until insulin was first isolated in the early twentieth century, most children lived only weeks or months after being diagnosed with diabetes. Medical insulin, first administered in Canada in 1922, was a miracle for these children, and is now generated synthetically. Today, the search for a cure continues. There has been some success with the transplantation of healthy, insulin-producing pancreatic tissue, so that insulin injections are no longer needed. But this is problematic because of the need for life-long immuno-suppressive drugs to prevent the rejection of foreign tissue. Because this 'cure' can be worse than the disease, transplants are only considered in type 1 diabetics when other treatments have failed.

Understanding and curbing the environmental factors that drive the disease would, of course, be a better outcome than curative strategies alone.

THE ROLE OF THE EARLY ENVIRONMENT IN TYPE 1 DIABETES:

All immune diseases basically occur as a result of inappropriate immune responses, either to specific self proteins (antigens) as in autoimmune conditions such as type 1 diabetes, rheumatoid arthritis, lupus, multiple sclerosis, and thyroiditis, or to environmental proteins (allergens) as in allergic diseases.[8] The pattern of disease is determined by the kind of antibodies generated, and which proteins they are directed against. Genetic predisposition plays a part in these conditions, sometimes evidenced by a family history, but the environment also plays a critical role.

The best evidence of an environmental effect is the rapid and parallel rise of so many of these diverse immune disorders within only a few decades.[9] In the case of type 1 diabetes (as with other immune diseases) there has been a progressive increase in many countries since the 1950s.[10] This strongly suggests that the immune system is specifically sensitive to modern lifestyle changes, and there has been speculation that this is related to impaired regulation of immune responses, such that rogue (autoimmune or allergic) responses are more likely to go unchecked or escape suppression.

The steepest rise in type 1 diabetes has occurred in *preschool* children, particularly in the last ten to twenty years,[11] indicating that the environmental exposures that drive this disease begin their effects early in life. In diabetes (as in allergic diseases) early differences in immune function precede the disease's development and can even be detected at birth, indicating environmental influence begins even before birth.

It is also likely that there are *initiating* factors that set off the abnormal immune response, and *precipitating* factors that determine if, when and how the disease is expressed. Not everyone with allergic antibodies or autoimmune antibodies will develop allergic or autoimmune disease. While they are clearly prone, other factors may determine if these potentially harmful immune responses translate into tissue damage and disease.

In diabetes, a long 'silent gap' between the initiating environmental exposure and onset of disease suggests that factors modulating the rate at which the disease unfolds may also be important. There is speculation, for example, that greater early-growth rates in the latest, larger generation increase the workload on islet cells and may promote earlier onset of disease and disease progression,[12] while other factors may initiate the original immune reactions.

THE 'HYGIENE HYPOTHESIS' AND IMMUNE DISEASE:

One theory behind the rising rate of immune diseases is that our progressively 'cleaner' environment (less rich in biodiversity such as fungi and bacteria) may not provide the required level of early immune stimulation for 'optimal' immune maturation (Chapter 10). This might explain why we are more prone to inappropriate immune responses against otherwise harmless proteins.

For diabetes, some global trends seem to support this theory. Rates of type 1 diabetes are highest in countries where exposure to infectious disease is lowest. A recent study found support for this theory when they related global diabetes to patterns of infectious diseases. The researchers found that people living in regions where the microbial burden was low were significantly more likely to develop the disease.[13] While this provides support for the hygiene hypothesis, it is only guilt by association, and does not prove causation.

The patterns and biodiversity of our gut bacteria are important in diabetes because of their dual role in maintaining both immune and metabolic balance. Children with type 1 diabetes have been noted to have increased gut permeability, intestinal inflammation and impaired immune regulatory mechanisms.[14] Studies show patterns of gut flora in type 1 diabetic children differ from those in healthy children with similar feeding history and genetic risk.[15] Diabetic children had much lower levels of the 'healthy' bifidobacteria, which enhances the intestinal epithelial barrier function and suppresses inflammation (Chapter 10). This adds to the growing circumstantial evidence.

In animal studies, factors can be kept constant, making them easier to assess. When animals are raised in a 'germ-free' environment, the immune system fails to develop the processes that suppress autoimmunity and allergy, and they quickly develop immune diseases.[16] Bacterial exposure can reverse this, but only at an early age. Again, we find a situation where the critical window of influence is in early development. In diabetes-prone animals, exposure to selected bacteria, viruses, parasites and even inactive microbial products can prevent autoimmune diabetes – providing hope that the same kinds of strategies might one day help protect or 'vaccinate' against type 1 diabetes in humans.[17] New studies reveal the importance of the interaction between microbes and the 'innate' microbial sensing parts of the immune system (Chapter 10) in preventing the onset of diabetes.[18] Because this is an early-onset disease, any interventions will have to be very early in life.

The 'hygiene hypothesis' still fits with the theory that infections might *trigger* the onset of the autoimmune process in type 1 diabetes. In other words, a virus might initiate the development of a clone of rogue autoimmune cells, but a less mature system in a 'cleaner' environment may *fail to eliminate* or suppress the abnormal response.

OTHER EARLY RISK FACTORS FOR TYPE 1 DIABETES:
The possible modern risk factors for type 1 diabetes include nutritional changes, modern pollutants, changes in physical activity, indoor living and other lifestyle behaviours. Because these are often interrelated it is difficult to pinpoint which factors might be having a meaningful effect. And it is likely that the effect is the result of a combination of these factors.

For example, dietary composition is a major determinant of gut microbiota patterns. So in addition to our 'cleaner' environments, our high-fat, low-fibre, high-sugar foods affect the microbial diversity in our guts and contribute to our increased proclivity for inflammation. There have also been studies linking specific nutrient factors such as wheat, cow's milk, fruits and omega-3 fatty acids with the risk of type 1 diabetes.[19] However, there are many inconsistencies between these studies. There are now several large-scale international studies underway to try to unravel these relationships, such as the 'TEDDY' study.[20] This 'observational' study is looking at both the genetic and the environmental factors that contribute to type 1 diabetes, using standardised methods. There are other large-scale 'interventional' studies looking at the effects of specific factors, such as infant formulas in the 'TRIGR' study.[21] This study has recruited over 5,000 infants in fifteen countries to see if weaning to an extensively hydrolysed formula in infancy will decrease the risk of type 1 diabetes later in childhood. A small pilot study suggested this reduced the risk of developing diabetes-associated auto-antibodies by between five and ten years,[22] although it will be some years before the results are available.

Because none of these potential risk factors act in isolation, we should study their interaction and not expect a 'one size fits all' strategy to prevent diabetes in children at risk. There are likely to be a range of genetic and environmental factors which vary between individuals. Even so, understanding how and why our modern lifestyle has produced a pandemic of childhood diabetes is an important step toward reversing it.

TYPE 2 DIABETES – A NEW DISEASE IN CHILDREN:

Until the 1990s type 2 diabetes was only seen in adults, typically after the age of forty. However, with increasing childhood obesity, in the USA alone between 2002 and 2005 there were 3,600 children and adolescents newly diagnosed with type 2 diabetes each year.[23] In children, the disease seems to progress and become resistant to treatment more quickly.[24] Children appear less likely to respond to lifestyle interventions and oral medications than adults, and may require insulin shots much earlier in their disease process. The reasons for this are not clear, but may be related to the rapid growth and the intense hormonal changes of puberty.

Type 2 diabetes basically occurs when the islet cells in the pancreas fail to meet an increased demand for insulin caused by 'insulin resistance' in the tissues. With over-nutrition and obesity, the adipose tissue releases increased amounts of fatty acids and inflammatory cytokines. The resulting inflammation, cellular stress and hormonal changes eventually lead to reduced sensitivity to the effects of insulin – or insulin resistance.[25] As the tissues become less responsive to insulin, the islet cells in the pancreas try to compensate by producing more and more insulin. Over time, this compensation fails as the pancreatic islet cell function declines.[26] In the early stages, lifestyle changes can reverse many of the metabolic changes of diabetes. The problem is that few people achieve the level of change that is needed. Once the disease becomes well established, and islet cells also begin to fail, this is not reversible. Even so, exercise and a healthy diet remain important for reducing progression.

There is a genetic predisposition and not everyone who is overweight or eats a lot of sugary foods will get this condition. Some people carry genetic variants (genetic polymorphisms) that make them more predisposed to diabetes when they have a risky lifestyle.

EARLY LIFE ORIGINS OF TYPE 2 DIABETES:

Interventions very early in life aimed at preventing type 2 diabetes are clearly important. Because this form of diabetes is the result of a *failure of the islet cells to produce sufficient insulin* in the face of insulin *resistance in the body tissues* we need to look in two directions for the early-life risk factors. First, factors that promote insulin resistance. Second, those that promote reduced islet cell capacity. Or both. Early-life factors that appear to increase both of these risks include over-nutrition or under-nutrition in early life, and exposure to high blood sugars in pregnancy from maternal diabetes.

Like other organ systems, the insulin-producing beta-islet cells are formed in foetal life and continue to grow and develop into the early postnatal period. The increased metabolic demands of high blood sugars in diabetic mothers can increase the islet beta-cell mass. This is an adaptive response in the short term, but can cause low blood sugars immediately after birth as maternal blood sugar is withdrawn. This also predisposes to eventual beta-islet cell failure and diabetes in later life – especially if there are additional postnatal stresses, such as overweight and insulin resistance.

In pregnancy, both over-nutrition and under-nutrition also increase the risk of diabetes. Again, this may be related to the risk of obesity and insulin resistance as well as effects on islet beta-cell development.

In animals, a range of restrictive dietary conditions in pregnancy promote type 2 diabetes in offspring, including caloric restriction or protein restriction of the mother, and other factors that impair placental function and reduce nutrient supply to the foetus. In humans we see similar patterns, with higher rates of type 2 diabetes, obesity and heart disease in people of low birth weight, reflecting reduced nutrient supply during their gestation.[27, 28] Studies show those born small are several times more likely to develop type 2 diabetes.[29] Because low birth weight is a risk factor for obesity (Chapter 4) it has been proposed that insulin-resistance (caused by obesity) may be the mechanism of this effect. However, it has also been proposed that under-nutrition during critical stages of gestation may lead to reduced beta-islet cell formation with lower functional reserve. Both of these are likely to be true. Children and adults who had low birth weights are more prone to obesity and insulin resistance, but they also appear more prone to impaired beta-cell function compared with those of normal birth weight.[30, 31] Furthermore, growth retarded human foetuses have reduced pancreatic islet cell mass compared with their well-nourished counterparts.[32]

At the other extreme, maternal over-nutrition and obesity also increase the risk of type 2 diabetes in offspring (Chapter 4). This is apparent in both human and animal models of obesity. High-fat diets in rat mothers appear to program for insulin resistance, increased body weight, high blood pressure, and the metabolic syndrome associated with type 2 diabetes.

Fathers also have a role to play. When *father* rats are fed on a long-term high-fat diet prior to conception, their offspring show increased body weight, adiposity, higher blood sugars and abnormal insulin responses.[33] This is a 'non-inherited' effect due to diet-induced alterations in gene expression through 'epigenetic' changes (Chapters 4 and 11). This adds further complexity to the intergenerational story.[34]

The effects of early-childhood over-nutrition are more obvious, particularly with increases in both obesity and type 2 diabetes in children. Here again, we have learned from animals how high-carbohydrate feeding, even when limited to only the suckling period, can induce adaptations in energy metabolism that persistent in adulthood, with obesity, higher blood sugar responses and glucose reduced insulin secretion.[35] As with other NCDs, the effects are even greater when *postnatal* over-feeding occurs following under-nutrition in the *prenatal* period.

As insulin resistance is closely linked with overweight, the early origins of type 2 diabetes closely parallels the obesity story with many of the same risk factors. The lifestyle behaviours and environmental stressors implicated in the obesity epidemic may also alter early beta-islet cells' growth and reduce the capacity to deal with metabolic stress. Other factors such as infections, mental stress and medications (such as steroids) can also predispose to type 2 diabetes without necessarily causing obesity as well.

GESTATIONAL DIABETES – FUELLING THE BURDEN OF DIABETES:
Diabetes is fast becoming one of the most common complications in pregnancy. This can be seen in most regions of the world, reflecting the increasing frequency of type 2 diabetes in most populations. Although 'gestational' (in pregnancy) diabetes is transient in most cases, it significantly increases the risk of future type 2 diabetes for both mother and child (below). The greatest upsurge in diabetes is in the developing world, where maternal education is more limited and healthcare resources have been channelled towards other priorities. In some regions gestational diabetes has increased by between 10 and 100 per cent over the last twenty years.[36] Countless cases still go unrecognised and untreated, fuelling the spiralling rates of diabetes and obesity across generations.

Gestational diabetes is triggered by the significant physiological and metabolic shifts made by the mother to cater for her new passenger. Pregnancy hormones and other factors interfere with the action of insulin. Obese and overweight women, and women with a family history of type 2 diabetes, are more prone to gestational diabetes. So are smokers and some racial groups, such as Pacific Islanders, indigenous Australians and North Americans, and African Americans. However, even women with few obvious risk factors can develop high blood glucose levels for the first time while pregnant. This is why all pregnancies should be routinely screened for gestational diabetes.

Babies born to mothers with gestational diabetes are at increased risk of being large for gestational age, with associated birthing difficulties. They are exposed to high blood-sugar levels during gestation, their insulin production is exaggerated to control this and their islet cells become relatively enlarged. After birth the high insulin levels can cause the baby's blood sugars to fall to dangerously low levels. If not treated, this can cause seizures and permanent brain damage. This is why these infants are fed early and their blood sugar levels are monitored closely.

These babies look 'big' but many of their organ systems are *less* mature. They are more prone to respiratory distress and jaundice due to immature lungs and liver function. They are also at risk of abnormalities of the heart muscle and metabolic effects such as low calcium and magnesium levels. High red blood-cell counts make the infant's blood more viscous and can increase the risk of strokes and other neurological defects. While these imminent risks are the most obvious and best studied, there are also long-term effects that increase the risk of future NCDs.

As *adults,* children of diabetic mothers are more likely to have abnormal islet insulin secretion, even if they have not developed diabetes themselves.[37] Together, genetic heritability and exposure to high blood sugar during gestation set the scene for a much higher risk of diabetes in the next generation. In this way, diabetes has potential to become more common with each generation.

For all these reasons, screening and treating diabetes in pregnancy is now being addressed through the new global priorities set to address NCDs by WHO and the United Nations' health agenda (Chapter 2).

However, we still need to understand the exact degree or level at which higher blood-sugar levels create future risk and when the most critical periods are. Animal studies suggest that there are sensitive periods beginning even before the embryo implants. In early gestation, adverse conditions as the pancreas starts to first differentiate can result in fewer islet cells. And later in gestation, when the islet cells are multiplying, adverse conditions can affect their blood supply (vascularisation) and growth (proliferation), with long-term effects. Even after birth, nutritional patterns can significantly influence the size and the function of the islet cell mass, increasing vulnerability to diabetes.[38] We still need to understand how this translates to humans to know how to best reduce the risk of future disease.

Losing 'old friends' – gut biodiversity and diabetes:
As we have already seen, friendly bacteria (our symbiotic microbiota) have many important metabolic and immune functions. Disruption to the delicate balance in our intestines is implicated in the rise in metabolic disorders (such as obesity and type 2 diabetes) and immune disorders (such as allergy, type 1 diabetes and other autoimmune diseases).

There is some evidence that gut microbiota have changed in Western countries with industrialisation.[39] These differences can be seen as early as the first week of life as our gut is first colonised – at a time when our gut immune system is undergoing the most critical periods of development. The gut contains the largest network of immune cells in the body. What happens (or fails to happen) in the gut during this early period of life has long-term implications for our immune health. This is why it is of particular interest that children with type 1 diabetes may have differences in their gut microbiota, immune regulatory mechanisms, and associated intestinal inflammation (as referenced earlier).

The 'gut microbiota' hypothesis also provides a link between type 1 and type 2 diabetes in the increased tendency for gut inflammation. This principle extends to other NCDs including allergic diseases (Chapter 10), mental ill health (Chapter 7) and obesity-associated NCDs (Chapter 4). It provides an attractive shared explanation for why so many of these inflammatory conditions have arisen under the same environmental changes.

Type 2 diabetes, like obesity, has associated chronic low-grade inflammation with increased levels of multiple inflammatory mediators. And, like obesity, type 2 diabetes has associated differences in gut colonisation patterns. However, preliminary evidence suggests type 2 diabetes may be associated with subtly different colonisation patterns to those in obese people who don't have diabetes.[40] This suggests that the microbial patterns that predispose to impaired glucose metabolism might be different from those that predispose to weight gain specifically. However, this is still a new field of research and things are far from certain.

Improving gut colonisation and nutrition in pregnancy might improve the metabolic and immune health of both the mother and her foetus. Increasing dietary fibre (prebiotics) and reducing fat intake is one way to promote favourable colonisation (Chapter 10). Another way is to give probiotic supplements. In a Finnish study, 256 pregnant women were randomised to one of three study groups in their first trimester. One group received a probiotic bacteria mix *and* healthy dietary intervention (to reduce fat and increase fibre), the second group received *only* the dietary intervention (but no probiotics) and a control group received *neither* the dietary intervention nor the probiotics. The mothers who received the probiotic mix were less likely to develop gestational diabetes than those on the diet only, or those in the control group.[41] There were other benefits in those who received the dietary intervention, with reduced risk of large birth size,[42] and improved metabolic health at six months of age.[43] These infants had improved insulin metabolism and this was favourably associated with reduced skinfold thickness and waist circumference.[44] There were even some effects on breastmilk composition, including the levels of hormones (adipokines) that regulate fat metabolism. The metabolic effects of a high-fibre diet are due, at least in part, to promoting favourable colonisation with bifidobacteria and many other friendly species. Changes in dietary composition can quickly regulate intestinal microbial composition and function.[45] This means that dietary strategies may ultimately be more effective than only taking a probiotic 'pill' containing a few bacterial strains (Chapter 10).

Although more studies are obviously needed, these findings support the concept that relatively simple and inexpensive strategies in early life can have positive metabolic effects on both the mother and foetus and reduce the risk factors for obesity, diabetes and potentially other future diseases.

RESTORING THE BALANCE:

Many of the 'risk factors' for diabetes are modifiable, giving us cause for optimism. It is likely that simple strategies to restore more traditional nutrition/colonisation patterns will have benefits on many organ systems. For example, there is already good evidence that prebiotic fibre, probiotics and other dietary strategies in pregnancy or lactation can reduce the risk of other NCDs such as eczema (Chapter 10).[46] And we have already seen how the potential immune benefits of reducing metabolic stress and inflammation in pregnancy may also reduce the risk of type 1 diabetes and other immune diseases.

Healthy dietary patterns also include higher levels of omega-3 fats, antioxidants, and other vitamins. Antioxidants can reduce the metabolic stress that predisposes to obesity and type 2 diabetes.

In animals, giving the mother antioxidants in pregnancy reduces oxidative (metabolic) stress and inflammation and prevents poor glucose tolerance (pre-diabetes) and adiposity in her adult off-spring.[47] Reducing oxidative stress is directly linked with our general life expectancy, and the length of the telomeres at the end of our chromosomes. These are our 'biological clock' and shorten as we get older (Chapter 11), but there are wide variations between individuals and with environmental exposures. It is well known that over-nutrition can tamper with our biological clock and reduce our life expectancy, and that reducing protein and calorie intake can improve it. But protein and calorie restriction in pregnancy and early development can have the opposite effect and actually reduce life expectancy.[48, 49] This happens through reducing antioxidant defences and telomere length.[50] So it is very important that our diets are both balanced and 'stage appropriate' (Chapter 11).

FATTY LIVER DISEASE: ANOTHER NEW DISEASE IN CHILDREN:
Non-alcoholic Fatty Liver Disease (NAFLD) is now the most common form of chronic liver disease in both adults and children living in industrialised countries such as the USA.[51] It is closely linked with the pandemic of obesity, metabolic syndrome and type 2 diabetes, and shares the same lifestyle risk factors. NAFLD condition was first identified in the late 1970s when over 20 per cent of an American population were discovered to have fatty liver changes similar to alcoholic liver disease, but without any history of significant alcohol consumption.[52-54] Almost 40 per cent of obese people show evidence of this disease – many without symptoms. In particular, consumption of sugary soft drinks and fructose-sweetened foods (below) is a major risk factor, particularly in children.

The first cases of NAFLD in children were not described until the early 1980s.[55] Now, around 10 per cent of all US children have

NAFLD with important ramifications for their long-term health; it affects more than 17 per cent of adolescents by nineteen years of age.[56] The rates are higher in Hispanic and Asian children, but lower in African Americans, highlighting genetic susceptibility. Thirty-eight per cent of obese US children have NAFLD.[57] This is not unique to the USA. A study of 966 Iranian school children found an overall NAFLD incidence of 7 per cent with higher rates (of 12.5 per cent) in older children.[58] They also found that children with NAFLD were four times more likely to have insulin resistance.

As the name implies, fatty liver disease develops with accumulation of fat in the liver tissue. This is called 'hepatic steatosis', and as the condition advances the liver becomes more inflamed with progressive tissue damage, scarring (fibrosis) and destruction. This is referred to as 'nonalcoholic steatohepatitis' (NASH) and, typically, 20 per cent of patients with NASH will reach cirrhosis and 'end-stage' liver disease within ten years.[59]

The liver is essential for breaking down by-products of metabolism and other substances, including drugs. It also produces many of the body's essential proteins, such as clotting factors and transport proteins, which carry hormones and nutrients around the body. As these functions fail, many systems are affected by the build-up of toxic factors, bruising and impaired clotting, hormone imbalances and disruption of protein transport systems. The blockage of vessels entering the liver leads to venous congestion in the abdomen and other abdominal organs, impairing their function as well. The abdomen progressively distends with fluid (ascites), which fills the abdominal cavity, while the rest of the body wastes away. The abdomen venous congestion (portal hypertension) causes varicose veins in dangerous places, such as inside the oesophagus where they are prone to catastrophic bleeding. There is also a risk of liver-cell cancer in patients who develop NASH. Any one of these changes can eventually lead to death.

This paints a dismal picture. Over 10 per cent of patients who develop NASH will die of liver-related death within a ten-year period. On the other hand, 'steatosis' alone is more benign and progression to cirrhosis only occurs in 3 per cent of these patients.[60] The major risk factors for *progressing* to fibrosis are obesity and diabetes, particularly in people over fifty years old. And the disease responds well, at least in the early stages, to the treatment of diabetes and weight control.

Because symptoms are uncommon and nonspecific, early diagnosis depends on a high level of suspicion and screening with ultrasounds and blood tests to assess liver function.[61] This, in the context of body weight and other factors, will determine the need for a liver biopsy which is ultimately the most accurate way of determining the presence and degree of disease.

EARLY-LIFE RISK FACTORS FOR FATTY LIVER DISEASE:

Once again, growth restriction in pregnancy and low birth weight appear to be early risk factors for fatty liver disease, especially if followed by rapid growth soon after birth. This is linked with both early obesity and type 2 diabetes. Nutrient deficiency during foetal life can increase expression of insulin receptors as an adaptive mechanism. This, in turn, can predispose to rapid weight gain and accumulation of fats in the liver and other body tissues post-birth. This makes the liver more vulnerable to secondary 'hits' such as inflammation, oxidative stress and toxins that lead to progressive liver damage. Again, disturbances in the gut microbiome might also contribute to this.

Childhood obesity and genetics are also risk factors, while breastfeeding can reduce the risk of metabolic conditions, including fatty liver.[62]

High consumption of soft drinks and processed foods containing 'added sugars' increases the risk of fatty liver, diabetes and obesity in children. Most processed foods contain considerable amounts of added sugar: table sugars, high fructose corn syrup (HFCS) or other fructose sweeteners, other syrups, honey, dextrose and dextrins. While there is still no definite proof of the effects, there is some evidence that drinking two sugar-sweetened beverages per day for six months can induce features of the metabolic syndrome and fatty liver.[63]

Data collected in the USA (2001–2004) revealed Americans ate an average of 22 teaspoons of added sugar each day. The recommended intake is for no more than 5 teaspoons for an average adult woman, and 9 teaspoons for an average adult man. Teenage boys were consuming an average of 34 teaspoons of sugar per day! Girls of the same age consumed 25 teaspoons. Even preschoolers (one to three years) consumed an average of 12 teaspoons of added sugar per day.[64] And this is by no means unique to the USA.

Collectively, there is evidence that regular consumption of sugar-sweetened beverages is related to the risk of diabetes, metabolic syndrome and cardiovascular disease in both adults and in children.

Sugary drinks may also influence *taste preferences* in later life. Another US study during the mid-1990s found that around 30 per cent of infants were given sweet drinks before four months of age, and 14 per cent were given these in the first month of life.[65] And over 70 per cent of foods contain added sugar. The increasing consumption of added sugar has been associated with a *decrease* in the intake of important nutrient and food groups. A growing proportion of children are showing insufficient intake of some vital nutrients such as calcium (Chapter 8).[66] Sugars that occur 'naturally' in healthy nutrient-dense foods, such as lactose in milk or fructose in fruits, are an important part of a normal diet and are much healthier than foods with 'added sugars'.[67]

The steep increase in added fructose in the form of corn syrup (HFCS) has become a particular focus of this concerning trend. Unlike sucrose, the fructose molecules in HFCS are not bound to a corresponding glucose molecule, so they are immediately absorbed in the body and utilised straight away. These metabolic differences have raised concerns that HFCS may be implicated specifically in the rise in obesity, diabetes and fatty liver disease.[68-70]

Fructose may also have effects on the gut microbiota.[71] It potentially increases absorption of toxic inflammatory bacterial products such as endotoxin which influence appetite and other metabolic processes. Early studies in rats suggested that fructose was more likely to increase triglyceride levels than an equivalent amount of glucose.[72] Similar observations were made when overweight adults were given either fructose or glucose and compared ten weeks later. Although both groups gained weight, only fructose was associated with an increase in the 'visceral' obesity (more toxic fatty deposits).[73] The group that received fructose also had increased cholesterol and insulin resistance compared with the groups that received glucose.

In the USA between 1977 and 1996 there was also a dramatic increase in consumption of 'fast foods' such as sugar-sweetened beverages, pizza, and salty snacks.[74] The greatest increase was in young adults and children.

There is still great hope, however, that we are entering an age of greater social and individual responsibility.

Simple strategies should be put in place now: encouraging healthy weight gain in pregnancy; avoiding rapid weight gain post-birth, including in small babies;[75] breastfeeding; and restricting sugary-food-and-drink intake in babies and children. These strategies could reduce the expression of liver disease many years later.

Once children *do* develop this disease, *gradual* weight reduction is also an important goal, avoiding fatty, sugary food. Fish oil derivative 'DHA' (docosahexanoic acid) and vitamin E may also improve some symptoms of fatty liver.[76, 77] Antioxidants such as vitamin E counteract the oxidative stress and free oxygen radicals thought to play an important role in the progression from simple fatty liver to steatohepatitis. Of course, it is far preferable to avert the development of this disease in the first place.

PROMOTING METABOLIC HEALTH:

Our metabolic system is a complex network of biochemical processes that occur throughout our body tissues. Organs such as our liver, pancreas, kidneys and even our adipose fat stores play key roles in orchestrating metabolic balance, and we know that their structure, size and even their function can be set very early during our development, largely before we are born. As discussed, the composition of our gut bacteria (microbiota) is extremely important for our metabolic health, and this, too, is established within the first few years of life, giving us the bacterial 'fingerprint' that we carry with us throughout life (Chapter 4 and 10). Breastfeeding and our early diet are important in determining the bacteria with which we develop this life-long relationship. During these early years we can also hope to establish healthy eating behaviour, and healthy patterns of physical activity.

Although our best chance of reducing metabolic diseases lies in providing 'optimal' early growth, nutrition, and gut health, we also need to be realistic. We must remember that some children are highly genetically prone. Improving early conditions and minimising ongoing risk factors may prevent the disease in some, but for others it may only delay the onset of disease. Delaying the onset of these diseases is still of great benefit for quality of life and life expectancy.

Type 1 diabetes is programmed in early life and, like some other metabolic conditions, is hard to prevent because we don't yet fully understand *how* this occurs. In my career I have seen several families with identical twins where one twin has type 1 diabetes while the other escaped. Genetically identical, and even similar environments. But there must have been subtle differences. Did one get have antibiotics for an infection at a critical time? Did one catch a virus the other didn't? We don't yet have the answers to these questions, but there is intensive research in this area.

Although genetics has a role here, the environment is the only explanation for the steep rise in disease in very tiny children. If we understand this we have the best chance of preventing it, and so it is imperative that our communities and our governments continue to invest in this quest of discovery.

6

The early-life origins of heart disease
and hypertension

Cardiovascular disease (CVD) has been the world number-one killer for decades. And although we are making small and certain steps towards fighting it, again our best hope of long-term success is to begin this battle in early life.

In 2008 the World Health Organization estimated that 30 per cent of *all* global deaths are from cardiovascular disease. Just as with other NCDs, there has been a dramatic and disproportionate rise in CVD in low- and middle-income countries, which now account for 80 per cent of all global CVD deaths.[1] Most cardiovascular disease can be prevented by targeting risk factors such as tobacco smoking, unhealthy diets and obesity, physical inactivity, hypertension, diabetes and raised cholesterol. Deaths from CVD in many high-income countries have actually fallen since the 1970s due to both lifestyle measures and medications, proving that something *can* be done.

My very first research project was in cardiovascular medicine, trying to understand why vegetarians have lower blood pressure and lower risk of heart disease. It was then that I had my grounding in the causes and consequences of high blood pressure (hypertension) as a major risk factor for heart attacks, strokes, aneurysms, chronic kidney disease and heart failure. I soon learned that relatively simple measures to reduce blood pressure, even slightly, can increase our longevity and our quality of life.

THE FIRST SIGNS OF CARDIOVASCULAR DISEASE APPEAR IN CHILDREN:

Although the symptoms of CVD might appear in adult life, early physiological and metabolic indicators can be detected from an early age. Many of the antecedent risk factors begin their effects even earlier, and some of this risk programmed before birth during foetal life.

Premature thickening and increased stiffness of the arteries can be detected in children non-invasively using ultrasound Doppler studies of the carotid artery in the neck. By around ten years of age, obese children already have greater carotid intima-media thickness (IMT) as a measure of the arterial wall dimensions. Compared to healthy children, they also showed increased arterial stiffness, higher blood pressure and enlargement of the heart.[2, 3] There is a subgroup of children that will require lipid-controlling medications before adulthood, because lifestyle changes have not occurred early enough.[4]

The 'fatty streaks' that predate 'atherosclerosis' (hardening of the arteries) also appear in childhood. These irregular yellow-white discolorations are visible to the naked eye on the inside of arteries. From the age of ten they are common on the inside of large vessels like the aorta. And by twenty they can appear inside the coronary arteries that supply blood to the heart muscle. Fatty streaks are actually aggregates of inflammatory immune cells stuffed full of cholesterol and aptly named 'foam cells'. So, this brings the immune system into the story once again – my personal fascination as an immunologist – and, with it, all of the early factors that could influence early immune development to promote inflammation of the blood vessels. Not all fatty streaks will develop into atherosclerosis. But recent studies done in conjunction with our medical school (UWA) have shown that a strikingly high proportion of Australian fourteen year olds already had higher blood pressure, higher waist circumference, unhealthy blood lipids (cholesterol and triglycerides), abnormal insulin and liver function, as well as higher inflammatory markers than other children, putting them at risk of future CVD.[5]

Many of the risk factors for CVD are shared with other common chronic diseases, and begin to have their effects of programming risk even before birth. Already by the early 1970s, Norwegian researchers had begun to recognise that people born in regions with high poverty and infant mortality rates had much greater risk of later death from atherosclerotic heart disease.[6] They also noted that later exposure to prosperity and affluence was also a prerequisite for the disease. Nearly ten years later, David Barker and his colleagues proposed that foetal under-nutrition leads to adaptations that reprogram hormonal and metabolic responses and lead to permanent changes in body structures and functions that increase the risk for coronary heart disease in later life. Since then, both animal models and human studies have shed light on how this might occur.

To set the scene for this, it is useful to first consider the physiological processes that lead to atherosclerosis and vascular diseases.

SOME BASICS OF BLOOD PRESSURE:

With every heartbeat our blood pressure rises to its maximum (the 'systolic' pressure) as the heart pumps blood into the arteries, and quickly falls to its resting minimum level (the 'diastolic' pressure) between each heartbeat. We normally measure blood pressure in the upper arm (at the level of the heart) while sitting, and express it as the systolic pressure/diastolic pressure in millimetres of mercury (mmHg). Our blood pressure varies from moment to moment with our emotions, our patterns of activity, our eating, our sleep and the time of day. There are also wide variations between individuals, according to age, sex and many other factors. 'Hypertension' for adults is defined when the blood pressure reaches 140/90 mmHg or greater. However, the risk of CVD increases progressively for each increment above 115/75 mmHg, and levels in the range of 120/80 and 139/89 mmHg have been defined as 'pre-hypertension'.[7]

The fluid mechanics that determine blood pressure include the *volume* of blood, the *resistance* in the peripheral circulation and the *viscosity* or thickness of the blood. If the kidneys retain more salt and fluid, the increased blood volume will increase the volume of each heartbeat. If the blood vessels are stiffer, the peripheral resistance can also increase blood pressure. If the blood is more viscous the blood pressure increases and there is also a greater risk of clotting. Aspirin 'thins the blood' and has anti-inflammatory properties even though it can actually increase blood pressure, so low-dose aspirin effectively prevents cardiovascular complications in patients with and without hypertension.

All of this is tightly regulated. Blood vessels contain muscles and nerves that control the 'tone' by the level of constriction or dilation. Pressure receptors called 'baroreceptors' monitor our blood pressure, making immediate adjustments as we change our movements. The most important are in the aorta and in the carotid arteries supplying our brain. If the blood pressure falls for any reason, such as sudden blood loss, there is reflex adjustment through the nervous system. Signals are relayed through our brainstem control centre at lightning speed to constrict the peripheral artery muscles and increase the force and the rate of heart contractions. This helps to maintain blood pressure and the crucial supply of blood to the brain.

The kidney and adrenal glands also play an important role in longer-term adjustment and control of blood pressure. When it senses reduced blood flow, the kidney produces a hormone ('angiotensin II') that constricts blood vessels and increases blood pressure. This hormone, also produced in other tissues, triggers the production of adrenal hormones (aldosterone) that promote salt (sodium) and fluid retention by the kidney.

Although the basic aspects of blood regulation have been deeply studied over many years, we are still not clear on how genetic and environmental factors interact to promote hypertension. And, as you keep hearing – it will vary for each of us!

JUST HOW MUCH OF A PROBLEM IS HIGH BLOOD PRESSURE?

High blood pressure is a big problem. The World Health Organization has identified hypertension as *the* leading cause of cardiovascular mortality. Not only does it increase the risk of coronary heart disease, stroke, aneurysms, pulmonary embolism and peripheral vascular disease – it is also a major risk factor for dementia, blindness and kidney failure.

More than a quarter of the world's adult population has hypertension; and, unless things change, 30 to 40 per cent of the children born today are likely to have hypertension later this century.[8, 9] Hypertension is common in both developed and developing regions, and at least two thirds of hypertensive people are in low- or middle-income countries. Even in wealthy countries, hypertension is more common in the underprivileged.

In high-income countries such as the USA around 30 per cent of the adult population has hypertension. The rates of hypertension are higher in some racial groups, and in African American adults the prevalence is *already* greater than 40 per cent. Although the prevalence of hypertension remained fairly constant from 1999 to 2008, there was increased awareness, medication and blood-pressure control over this time. This may have contributed to declining death rates from CVD, although the nonfatal burden of disease remains high.[10, 11]

More than 50 per cent of people with hypertension don't even know they have it.[12] In China the awareness is less than 40 per cent, only 17 per cent are treated, and only 3.5 per cent adequately controlled.[13] Even in the USA only half of the people who *do* know that they have hypertension adequately control it.[14] Global awareness campaigns such as World Hypertension Day are designed to address this – an example of how organisations and media can get together to make a difference.

ECONOMIC IMPACT AND HEALTH COST BURDEN

The current global economic burden of hypertension and CVD is unparalleled, and is only likely to increase. In the United States alone, the *direct* medical cost of hypertension was recently estimated at $69.9 billion, and indirect costs, such as lost productivity, at $93.5 billion. Including conditions that hypertension causes (such as coronary heart disease, heart failure, and stroke), the estimated annual cost of hypertension is closer to $156.1 billion. In 2010, the cost of *all forms* of CVD was estimated at $444.2 billion, with hypertension as the most expensive component.[15]

Unless something is done, this going to get a great deal worse. Between 2010 and 2030, the direct and indirect costs of CVD are projected to be *in excess of $1 trillion.*[16] The story is very much the same in many other developed regions. In the European Union direct and indirect costs of CVD are currently estimated at nearly €196 billion a year.[17] For many of the poorer areas of the world the human cost is far greater and much harder to quantify.

Treatment is expensive. It is not always effective. Obstacles to achieving blood-pressure control include adhering to medicine schedules and making lifestyle changes. Prevention is more cost effective in the end. But it requires a long-term vision and long-term investment. Even a small reduction in the prevalence of these common diseases will save millions of lives and billions of dollars.

THE CAUSES OF HYPERTENSION – SOLVING ANOTHER MYSTERY?

Once again, while genetics are a clear factor, they cannot explain the progressive increase in hypertension in so many regions of the world. And, once again, our genetic susceptibility appears more likely to manifest in a Western environment.

Researchers have discovered some genes linked to hypertension, but the results have not provided all the answers. There are actually quite a *large* number of very common genes in the population that have been linked to blood pressure, but they only explain a relatively *small* degree of the variation.[18, 19] Although some genes can have large effects on blood pressure, these have turned out to be very rare.

The physiological processes that cause our blood pressure's 'set point' to rise tend to happen with age — but this rise is more exaggerated in some of us. So is this just a matter of a shift in balance of our 'normal' regulatory processes? And do environmental factors just tip that balance?

A number of mechanisms have been proposed to explain the chronic rise in peripheral vascular resistance. Although there are some obvious culprits, the relative contribution of these is debated. Disturbances in renal salt and water handling (the angiotensin–aldosterone system) and increased activation of the autonomic nervous system are both implicated.[20, 21] Inflammation and blood-vessel damage also appear to contribute to increased peripheral resistance in hypertension.[22] There is a growing awareness of the role of the immune system in the development of hypertension, and it is proposed that inflammation and oxidative stress play a synergistic role in the development of hypertension.[23] It is likely that several of these physiological mechanisms act together.

Some of the environmental risk factors for hypertension may be tipping the balance in favour of higher levels of salt retention, nervous-system activation, or inflammation. These include high-salt diets, stress and anxiety, obesity, diabetes, and lack of exercise.

Although caffeine consumption and vitamin D deficiency have also been implicated, their role in hypertension is less clear. Coffee raises blood pressure in the short-term, but not over the longer-term, nor does it increase the risk of CVD in hypertensive people who regularly drink it.[24] The story with vitamin D is an interesting one. Animal studies have also shown vitamin D is integrally related to blood pressure. Disappointingly, so far the results of randomised trials have not shown consistent benefits of vitamin D supplementation.[25] However, it is possible that some of the protective effects of sunlight exposure (in moderation) could be due to other factors such as ultraviolet, which may modulate immune function and potentially protect from many inflammatory diseases.[26] This is still an area of great interest and intense research.

The role of stress is also still debated. Combining data from available studies, there is some evidence that mediation reduces blood pressure, and that yoga can reduce stress levels.[27] However, because of the lack of 'hard science', these techniques are generally overlooked, despite the many benefits they might have and the obvious lack of side effects. This is another area where more research is needed.

Additional factors which might promote vascular inflammation include tobacco smoking, air pollution, excessive alcohol consumption, and high sugar and fat consumption. Because of the close association between hypertension and heart disease it can be hard to determine which factors are promoting atherosclerosis and which are increasing blood pressure through other mechanisms. Addressing dietary and lifestyle risk factors can improve blood pressure and decrease the associated cardiovascular complications.

The lifestyle changes that *have* been shown to lower blood pressure include weight loss, exercise, reducing salt intake, increasing fruit and vegetable intake (Mediterranean-style diet), and reducing alcohol consumption,[28-33] although some effects may not be sustained in the long term.[34] The causes and the role of these factors are still debated.

WHEN WE ARE YOUNG; THE ORIGINS OF HYPERTENSION:

Again, this story begins before we are born. And so we need to start with maternal and paternal factors at conception, the environment and exposures during pregnancy, and our early infant environment.

Low birth weight, maternal smoking in pregnancy and other causes of foetal growth restriction, and rapid 'catch-up' growth after birth are all factors in the development of hypertension.[35-37] There are also factors that protect from hypertension such as breastfeeding.[38] Maternal factors at conception, such as age, also appear important,[39, 40] probably because older mothers are more likely to have smaller first babies. But fathers matter, too, and there is emerging evidence that unhealthy diets in fathers influence the metabolic health of the offspring, through 'epigenetic' effects (Chapter 11).[41]

A number of significant studies have seen that foetal exposure has effects on blood pressure in the first five to ten years of life. These studies, including the Western Australian 'Raine' Pregnancy Study, showed that children have detectably higher blood pressure by the age of five years if their mothers smoked in pregnancy, and this shows even as early as one year of age.[42] Blood pressure was also higher with lower birth weights, in non-breastfeed infants and in children who had become overweight.[43, 44] Because childhood blood pressure 'tracks' into adulthood, these researchers suggest interventions aimed at quitting smoking, breastfeeding, quitting smoking, and the prevention of obesity may reduce the 'population distribution' of blood pressure.

David Barker and collaborators in Finland found that people born with smaller placentas (reflecting poor nutrient supply) were also at much higher risk of developing chronic heart failure in later life.[45] And, again, the risk was even greater in those who showed rapid body-weight gain in early childhood.

One way in which foetal growth restriction may lead to hypertension is through effects on the developing kidney.[46, 47] Brenner and his team proposed that nutritional restriction may affect the formation of the 'nephron' filtration units that comprise this organ,[48] as low birth weight babies suffering from foetal growth retardation tend to have reduced nephron numbers.[49, 50] Rats with small-birth-weight pups (following maternal malnutrition) develop hypertension and salt sensitivity when they are adults.[51] They have smaller kidneys and increased expression of kidney angiotensin II that increases blood pressure.[52, 53] There is also some evidence that humans with hypertension may also have lower nephron numbers.[54, 55]

Animal models have since revealed that maternal malnutrition also leads to *inflammation* in the kidney, and that blocking this with anti-inflammatory drugs prevents the hypertension.[56] Invasion by immune cells, including T cells and macrophages (Chapter 10), appears to damage to small vessels in the kidney, activating several processes (including the expression of angiotensin II) that can increase blood pressure.

Maternal stress hormones (cortisol) can also change the responsiveness of blood vessels, and can reprogram the baby's long-term stress response. Researchers gave a synthetic cortisol to pregnant sheep and showed that their offspring had fewer nephrons and developed hypertension five years later.[57] Other foetal stressors, such as reduced oxygen levels, may also change the long-term cardiovascular function and blood pressure in offspring.[58] In animals this can change heart muscle development and cause increased susceptibility to abnormal cardiovascular responses as adults. The relevance to humans is not at all clear, although maternal obesity can lead to low oxygen conditions in the placenta, and can also influence the risk of later hypertension.[59]

Finally, and a recurring theme, the early influence of the immune system and inflammation has recently emerged as another important early life factor in the programming of atherosclerosis, and this is discussed in more detail below.

We can already see a number of early factors that might alter the 'trajectory' of our blood pressure appear quite early on. If we can understand this more fully, it may offer the best opportunities for steering things in the right direction from an early age.

ATHEROSCLEROSIS IS ANOTHER INFLAMMATORY DISEASE:

Although there are other forms of heart diseases, here we will focus on 'atherosclerotic' CVD as the major cause of heart attacks, stroke and death. Previously, this form of heart disease was thought to be due to fatty accumulation, but it is now recognised as an inflammatory disease. In fact, the immune system plays a fundamental role in driving all stages of atherosclerosis.

The fat-filled 'foam cells' in fatty streaks are actually scavenger immune cells called 'macrophages'. Other immune cells, such as T cells (Chapter 10), also accumulate in these fatty streaks and together they drive chronic low-grade inflammation. Over time, this can lead to the formation of atherosclerotic plaques and scar tissues that thicken, narrowing the blood vessels and eventually limiting blood flow.

Although the accumulation of immune cells and fatty deposits begins in childhood, it is still not clear what initiates this inflammatory condition in the first place. Over time arteries thicken with the build up of fatty materials such as triglycerides and cholesterol – in particular the low-density lipoprotein (LDL) or 'bad' cholesterol. There is gradual and progressive immune destruction of the lining of the arteries, leading to the formation of plaques containing excess fat and scar tissue. Cholesterol crystals form under the plaques and calcification occurs in older, more advanced lesions. Significant narrowing or 'stenosis' of the artery is a much later event, and may not occur in all people; this appears to be the result of repeated rupture of the plaques and the ensuing healing process.

Symptoms occur when blood flow is obstructed. And this can be a very sudden event. With fatal results. In some people it may the *first* and *only* sign of disease. Clotting or 'thrombosis' is more likely to occur in abnormal and inflamed vessels. This can block blood flow, and the loss of oxygen supply will quickly lead to death of the affected tissue. The effects can be catastrophic – a myocardial infarction (heart attack) when the coronary arteries are occluded, or a stroke when the cerebral arteries are occluded. Plaques can also break off into the blood stream to lodge in distant tissues also causing sudden occlusion. Atherosclerosis can increase the risk of aneurysms – ballooning of the weakened vessel, increasing the risk of rupture. If this occurs in large vessels like the aorta, there is little chance of survival.

So if inflammation of the blood vessels is such a major factor in this condition, how and why does this start? As with all NCDs, there is complex interplay between environmental factors (especially diet and lifestyle patterns), genetic risk and early programming.

ATHEROSCLEROSIS IN EGYPTIAN MUMMIES:
Despite the fact that the atheromatous plaques contain fat and cholesterol it is still unclear if they are an underlying *cause* of atherosclerosis. CVD is certainly more common in populations that consume high-fat diets, but atherosclerosis also occurs in traditional societies.

We now know it was common 4,000 years ago in Egyptian mummies, challenging the idea that this is just a 'modern problem' caused by smoking, fatty foods and lack of exercise. In 2013, a team analysed whole body CT scans of 137 mummies from four different regions including ancient Egypt and ancient Peru.[60] They were able to detect atherosclerosis as calcified plaques in the walls of arteries (definite atherosclerosis) or calcifications seen along the expected course of an artery (probable atherosclerosis). This revealed that just over one third of the mummies had 'probable' or 'definite' atherosclerosis, demonstrating that modern lifestyle and high-fat diets are not the only factor in the rapidly increasing rates of CVD in the developing regions of the present day.

CHOLESTEROL AT ITS WORST:

There are situations where high cholesterol levels are more directly linked to the early development of atherosclerosis. For example, familial hypercholesterolaemia, or 'FH', is an inherited condition, commonly due to a mutation in the LDL receptor gene that causes very high blood levels of the 'bad' LDL cholesterol. This greatly accelerates the inflammatory process that leads to atherosclerosis and much earlier onset CVD. Children who inherit *two* mutant genes show very early and very severe cardiovascular disease. There is visible accumulation of cholesterol-laden foam cells in the skin, sometimes from birth. Yellowish lumps and nodules appear over joints, around the eyes and various other parts of the body.

These children experience the symptoms of coronary artery disease from a very early age and sadly often die of heart attacks before their teens. Although this severe (homozygous) form is rare and only affects around one in every million children, it shows us that cholesterol levels are very important in the development and progression of disease. A much more common and less severe form of the disease occurs when only *one* mutant gene is inherited (heterozygous). This affects as many as one in 500 people, many without realising they have it. It also leads to premature onset CVD with symptoms by the age of 30 or 40, highlighting the importance of early screening and early intervention. There also appear to be other genetic factors that increase the risk of hypertension, atherosclerosis and heart disease in some families.

INFLAMMATION TAKES CENTRE STAGE:

Regardless of the genetic or environmental risk factors, inflammation appears to be the main driving factor. One of the best-studied 'markers' of the level of inflammation is the C-reactive protein or 'CRP'. This can be easily detected in the blood stream and is a good indicator of the general level of systemic inflammation at any moment. CRP is produced by the liver in response to inflammatory warning signals (the interleukin-6 cytokine) produced by immune cells (macrophages) or fat cells (adipocytes). Normally present in low levels in the circulation (usually less than 10 mg/L), CRP levels rise acutely during any form of infection, inflammation or tissue damage, sometimes many thousand-fold above base-line levels.[61] In this setting CRP is acting as part of our body's complex network of defence mechanisms. As the acute inflammation is treated, or settles naturally, levels fall quickly back to baseline levels. Although CRP does not give any indication of the site or the cause of the inflammation, it is used commonly used in the clinical setting to monitor the progress of an acute illness. However, more recently it was recognised that variations in the *baseline* level of CRP are also an important indicator of more chronic, low-grade inflammation and that this might be a useful measure of disease risk. It is interesting that the statin drugs that are used to reduce the risk of cardiovascular disease act to reduce both cholesterol levels *and* inflammation, with measurable reductions in CRP.[62]

A number of studies have now shown that people with elevated CRP at baseline are more likely to develop CVD,[63] type 2 diabetes,[64] and have higher 'all cause' mortality.[65] Based on these studies there has been some consensus that a baseline CRP of > 3 mg/L can identify individuals at high risk of cardiovascular disease.[66] About one third of adults in high-income regions of Europe and North America would fit into this category.

There is some evidence that CRP can promote secretion of inflammatory mediators by the vascular cells in the vessel wall[67, 68] and coats the LDL-cholesterol for increased uptake by the macrophages in atherosclerotic plaque.[69] These observations suggest that CRP could be directly implicated in the development of atherosclerotic lesions, even though it is also clearly a measure of other inflammatory factors.

There are significant differences in the CRP levels of populations around the world. Thomas McDade and his team of biological anthropologists have made close study of this. In developing regions of Asia and South America where lifestyle conditions are still quite traditional, CRP is significantly lower than in high-income regions such as the United States.[70, 71] This suggests that affluence is a risk for the kind of chronic 'silent' low-grade inflammation that leads to CVD and other NCDs. There are some 'pro-inflammatory' factors that undoubtedly play a role in CVD. Smoking is an obvious one.

SMOKING AND DIET

Smokers have higher levels of systemic inflammation than non-smokers. Indeed, we see that smokers have significantly higher CRP levels; but, importantly, people who give up smoking can achieve similar CRP levels to non-smokers.[72] Studies have found that CRP concentrations *can decline more rapidly* when *younger* men and women stop smoking.[73] Of major concern, significant elevations in CRP levels are present in 13- to 16-year-old smokers:[74] damage related to cigarette smoking begins soon after taking up. Over time, the chronic inflammatory effects of smoking and the direct action of tobacco products on blood vessels, greatly increases the risk of atherosclerosis and cardiovascular death. If only we could eradicate smoking we would prevent millions of premature deaths and save the global economy billions of dollars, within a single generation.

Dietary patterns are also of major importance and many nutrients could be implicated in the immune (inflammatory) and metabolic aspects of atherosclerosis. Recent studies have confirmed people who have a diet poor in fatty fish (omega-3 polyunsaturated fats) and rich in other meats are more likely to show elevated CRP, and exhibited pro-inflammatory characteristics twelve years later. Diets rich in vegetables and associated high intakes of antioxidant micronutrients and essential fatty acids help prevent elevated CRP.[75] But this area remains far from certain. Studies have shown that fish oil supplementation reduced CRP,[76, 77] and although omega-3 PUFA have anti-inflammatory effects, and can reduce cholesterol, their specific long-term effects on reducing CVD risk are controversial. People carrying some genetic polymorphisms appear prone to a proinflammatory response favouring atherosclerotic vascular damage[78] and may benefit more than others from a preventive diet. Data from multiple studies of nearly 40,000 volunteers found omega-3 PUFA supplementation for at least a year reduced the risk of cardiovascular deaths, all-cause mortality, and nonfatal cardiovascular events in at-risk patients.[79] However, reports from even larger analyses have failed to show a clear benefit.[80] While it is not a magic bullet, it has few, if any, significant side effects and *'the potential of using n-3 PUFA for prevention and treatment of heart disease should not be overlooked'*.[81] Prevention methods for any NCD are always likely to be more effective when they encompass many healthy lifestyle choices.

Some of the cardiovascular risk of obesity may also be explained by chronic inflammation. Adipose fat tissues express and release the pro-inflammatory cytokine 'interleukin 6', and appear to add to or even drive low-grade systemic inflammation. Indeed, CRP levels are significantly higher in overweight and obese people. This is seen even in younger adults (seventeen to forty) and when all other obvious causes of inflammation are excluded or accounted for.[82] Together with the higher cholesterol levels often associated with obesity, this 'inflammatory state' may promote the processes that drive atherosclerosis. Another explanation is that the 'obesogenic' diet may be having other direct pro-inflammatory effects through changes in the gut microbiota. In animal models we have already seen (Chapter 4) how a high fat, low fibre diet increase levels of bacterial endotoxins reaching the circulation, and increasing interleukin 6 and other measures of systemic inflammation.

MICROBES AND ATHEROSCLEROSIS: FRIENDS OR FOE?

As the role of inflammation in atherosclerosis and CVD became more obvious, so did interest in the role of bacteria and viruses: the most common triggers of inflammation. Initially, the focus was on the potential harmful effects of microbes in the genesis of the disease, and more recently an interest has developed in the potential protective anti-inflammatory effects of some microbes.

For over 100 years there has been speculation that infections may either cause or promote atherosclerosis. Because infection is a common cause of higher CRP, there was renewed interest in this when the association between increased CRP and atherosclerosis was recently discovered.[83] New technologies allowed scientists to detect bacterial DNA in plaques from atheromatous blood vessels. This revealed a large number of species of bacteria are present in these lesions compared with healthy blood vessels.[84, 85] Some bacterial profiles look similar to the patterns seen in dental plaque, suggesting they may have seeded from the oral cavity. Others appear to have come from the lower gut. If blood vessels are abnormal, bacteria are more likely to seed onto the vessel wall. This will activate the scavenger immune cells that reside in the plaques, and add to the inflammation in these 'foamy' aggregates. At present it is thought that bacterial colonisation is a secondary factor that promotes progression of atheroma development in the vessel wall.[86] Antibiotics do not seem to prevent CVD, however, and 'germ-free' animals will still develop CVD if they have other inflammatory risk factors.

Other potentially harmful bacteria and viruses have been implicated in CVD, although none of these associations has been proven. One such culprit, debated for some years, is *Helicobacter pylori*. Known for its role in causing stomach ulcers, there have been some studies associating H. pylori with a higher risk of coronary heart disease, unhealthy cholesterol levels, and increased inflammation. There are studies in support of this case, but almost an equal number arguing against it.[87] And it is much the same story with a number of other bacteria and viruses that have been isolated from plaques.[88]

Given that foam cells and fatty aggregates appear in blood vessels very early in life, it is not clear what role bacteria might play in the earlier stages of this process. Even though the exact role of bacteria in the development of CVD is still not completely clear, it has certainly not been ruled out.

Although more investigation is needed to confirm this, studies suggest that a lower fat, higher fibre diet (promoting 'friendly' gut bacteria) may protect from CVD and metabolic syndrome through a broad range of mechanisms – including lowered cholesterol and inflammation.

As well as playing an important role in regulating our energy and fat metabolism, microbes in gut bacteria produce a range of metabolic by-products. Although this microbial digestion is generally helpful, not all of the metabolic by-products may be beneficial. Recently, several of these 'metabolites' produced by bacteria have been linked directly to CVD.[89, 90] But this may also be due to the food that we give them to digest.

A group of researchers in Cleveland, Ohio first identified metabolic products in higher levels in the blood of people with atherosclerosis.[91] These turned out to be digestion products of phosphatidylcholine, a dietary fat found in many foods including meats, eggs and dairy foods. They also showed in animal studies that gut bacteria are essential for the digestion and production of these metabolites, and that metabolite levels could be manipulated by changing the level of bacteria in the gut. The levels of metabolites reaching the blood stream directly influenced the function of the macrophage immune cells in the blood vessels, augmenting cholesterol accumulation and foam-cell formation to promote atherosclerosis.[92] Other studies, also in mice, have shown that adding 'friendly' probiotic bacteria can decrease the levels of these metabolites.[93] The relative importance of these observations in humans is not really known yet. But it demonstrates that bacterial and dietary factors can alter the expression of our genes to alter our susceptibility to disease.

Atherosclerosis, like so many NCDs, is a very complex interplay between our genes, our bacteria, our diet, our wider environment and our lifestyle. Ninety-nine per cent of the genes in our bodies are actually the genes of the bacteria that live inside us, so it is no wonder that they can have powerful effects on our metabolic and immune functions.

Bearing in mind that the life-long patterns of our microbiota are set *in the first years of our life*, we can again see the enormous potential for early-life factors to shape the bacterial patterns that in turn shape the metabolic patterns which influence our long term health.

FIGHTING INFLAMMATION WITH SLEEP AND EXERCISE:

Exercise can reduce our cardiovascular risk. As well as burning calories and improving circulation, physical activity reduces inflammation. Regular exercise reduces the levels of inflammatory cytokines in the circulation. And even in patients with established coronary artery disease exercise interventions for at least two weeks show a consistent and significant reduction in CRP and other inflammatory markers such as Interleukin-6.[94] This was seen in analysis of data from over twenty studies.

And then there is sleep. For some years there has been a recognised association between sleep disturbance and cardiovascular disease. Increased inflammation may be one reason for this. Sleep deprivation and sleep restriction increase CRP. In studies of people kept awake for 88 hours straight, CRP concentrations increased steadily and significantly each day and remained high during the recovery phase.[95] Blood pressure also increased significantly. In another experiment, volunteers slept a maximum of 4.2 hours each night, for ten consecutive nights.[96] They showed a four-fold rise in their CRP over the period. The researchers proposed that sleep loss may activate inflammatory processes and contribute to the increased cardiovascular disease reported with chronic sleep disturbance.[97-99]

Snoring and disturbed sleep is not uncommon in young children. And this risk factor is often overlooked. Enlarged tonsils and adenoids are a common cause of 'obstructive sleep apnoea'. Repetitive pauses in breathing (apnoea) can last more than 20 seconds, causing oxygen levels in the blood to fall. During the day the child may be 'overtired', or have learning and behavioural problems including hyperactivity, attention deficit, and poor memory. These children also have significantly higher CRP and inflammatory markers produced by immune cells such as Interleukin-6.[100, 101]

For every individual who develops CVD there will be a range of risk factors, genetic and environmental, acting together to drive the disease. Modern medicine cannot yet fully explain atherosclerosis or how it develops. But we *can* say that inflammation is a common feature in all patients with atherosclerosis, and that *many* CVD risk factors promote inflammation. It is also increasingly evident that many of these risk factors begin early in life.

ARE THE 'DYNAMICS OF INFLAMMATION' PROGRAMMED EARLY IN LIFE?

The most common and striking cause of acute inflammation at any age is infection. Bacterial infections, in particular, cause a sudden spike in inflammation markers such as CRP. So why are CRP levels *lower* in poorer countries such as the Philippines and Ecuador, where infectious diseases are much more common than they are in the USA?[102] This is what Thomas McDade and his team set out to discover, using data collected from these populations over several decades.

At first, they thought that this might just be because people were thinner in the Philippines and Ecuador. But when they compared people of the *same* waist circumference or skin-fold thickness, CRP levels were *still much lower* than in adults from the USA. If anything, their findings suggested that the capacity of body fat to promote inflammation varies between populations.

Genetic differences are a source of some variation in CRP within and between populations, but genetic background only contributed slightly to the differences in CRP between adults in the USA and the Philippines.[103]

The answer was clearly in the environment. And the environments could not have been more different.

McDade and his colleagues started to ask some very interesting questions. If inflammation is an immune response, could there be something different about the early immune development between these populations? Could early infectious exposures have lasting effects on the regulation of inflammation in adulthood? Could better-developed immune regulation dampen or prevent unnecessary inflammatory responses later in life? And could this tie into the 'hygiene hypothesis' proposed to explain rates of other inflammatory diseases such as allergy (Chapter 10) and type 1 diabetes (Chapter 5) in the West?

When these comparative studies began in the early 1980s, Filipino families lived in settlements ranging from remote rural areas and squatter camps with no electricity or running water, to more affluent urban neighbourhoods.[104] More than half had chickens, goats, or pigs roaming around the house, with very high levels of microbial exposure. Infections and diarrhoea were very common in infancy.[105, 106] This was the perfect opportunity to test their hypothesis in a population with a similar genetic background.

The differences were substantial. Children exposed to animal faeces and diarrhoeal infections in infancy had significantly lower CRP levels as young adults.[107] Children born in the dry season (with higher rates of infection) also had a much lower CRP by adulthood. This seemed consistent with evidence that early bacterial exposure promotes immune regulation and reduces the risk of inappropriate immune responses, such as allergies, later in life.

More frequent and diverse microbial exposures with more frequent bouts of acute inflammation appear to lead to repeated activation and de-activation of inflammation. This dynamic promotes more competent regulatory pathways, which can more effectively switch off inflammation when it is not needed. Inflammation has evolved as an essential part of immune defence. The regulatory pathways that switch *off* the immune responses once the threat is over are *equally important*. Less-competent pathways result in a greater propensity for chronic low-grade inflammation, and more sustained inflammatory responses to a triggering event in later life. *'Natural selection could not anticipate the highly sanitized, low-infectious disease environments currently inhabited by humans in affluent industrialized settings, and a poorly educated immune system may be the result'.*[108]

It appears to be the chronic low-level inflammation, rather than acute 'spikes', that are associated with CVD.[109] This needs to be investigated further. And we know it is not likely to be the only explanation. In the meantime, it seems reasonable to consider that the global rise in CVD and diabetes is not purely due to nutritional changes, but that lifestyle changes which have reduced microbial exposure to levels not experienced previously in the history of the human species may also have some part to play.

Microbes have well-known effects on early immune development, but so do many other things. Nutrition, pollutants and other common exposures have well-known effects. Many of these effects begin in pregnancy when immune responses also start to develop. In fact, we can already detect effects in cord blood at birth.[110-112]

Maternal smoking, dietary patterns and microbial exposure in pregnancy are also implicated in later onset inflammatory disease such as CVD.[113] Low birth weight is used as an indicator of a sub-optimal prenatal environment, and has been linked with a higher CRP later in life in a number of studies.[114-116] Recent studies show that early poverty is also associated with higher CRP,[117] and exaggerated inflammatory responses in adolescence or adulthood.[118, 119] Finally, the protective effects of breastfeeding against obesity[120] and cardiovascular risk may be the results of both metabolic benefits and anti-inflammatory effects.

A range of early microbial, nutritional and toxic exposures appears to 'program' or 'condition' the level and patterns of inflammatory response as we grow and age. Exactly how this manifests in each of us will depend on our ongoing environment and our genetic predisposition.

The link between the early environment and CRP fits very well with what we already know about CRP and cardiovascular risk. It also fits with the long-observed associations that early growth and development patterns are important determinants of later cardiovascular disease. All of these observations appear to stress the importance of the early environment for *both* metabolic and immune programming. And given that it is almost certain that atherosclerosis is driven by a *combination* of metabolic and immune processes, we would be foolish not to further explore how these pathways might become established very early in life, and how they interact during this early period.

CARDIOVASCULAR DISEASE AND THE INDIGENOUS HEALTH DIVIDE: The terrible longevity inequity between Australian Aboriginal and Torres Strait Islander people (with a life expectancy of only 56·3 years for men and 62·8 years for women)[121] and the Australian average (almost twenty years longer) is largely due to chronic NCDs, in particular cardiovascular disease and diabetes. The risk factors and trajectory for later ill health clearly come into play very early.

Nearly twice as many low-birth-weight infants (less than 2,500 g) are born to Australian Aboriginal and Torres Strait Islander mothers; and teenage pregnancy, a major risk for low birth weight, is *five times* higher in Indigenous populations.[122] These children *start* life at a disadvantage, and their risk only becomes amplified with age as they are also more likely to experience other risk factors for CVD and diabetes through ongoing social and economic disadvantage.

The higher rate of infant infection seen in this disadvantaged population does not seem to protect from later inflammatory diseases. Is this because of genetic differences? Or is this because the many other risk factors for inflammation and NCDs are so much stronger? We don't have the answers, but there is little doubt that to 'close the gap' we need to understand the deeper complexities and the social determinants of health.[123]

These disparities reveal that social determinants are primary drivers of biological risk. Recognising this, the World Health Organization established a Commission on Social Determinants of Health, tasked with linking knowledge with action through social policies beyond the health sector: '*There is no choice. If the major determinants of health are social, so must be the remedies*'.[124] For the indigenous populations of North America and New Zealand there has been significant progress in 'closing the gap', which proves that this is possible (Chapter 12).

• • •

Most doctors, me included, probably do not measure blood pressure or other early risk factors as often as we should in young people. Later, we *can* address hypertension and other risk factors with lifestyle change and medications as needed. But *if* there were earlier opportunities, and *if* earlier life conditions were more conducive to healthy choices, hypertension and atherosclerosis might be avoided in a substantial number of people.

Understanding the *biology of risk* is one thing. Taking action to overcome the many adverse factors that limit our health and prosperity is quite another. This will require a richer understanding of the factors that drive inequity and disadvantage. It raises many public-health issues around the broader *social determinants* of disease that govern our choices and our opportunities at all stages of life.

7

Brain, behaviour and mood

Our whole world depends on what is going on inside our head. How we see the world. How we react to it. How we feel. How we perform in society, at work, and in our relationships. How we adapt and how we cope. We are quickly defined by our ability and performance from an early age. Our future income and social status are heavily determined by our academic abilities and success in modern educational systems.

The importance of the early life for learning and development is an 'easy sell'. Parents, communities and governments all *know* this is important. Most strive to provide every opportunity they can, even in the face of limited resources and social inequity. Most dream that their children will have a richer and more prosperous adulthood than they have. And some invest everything they have for this.

As a young paediatrician I often assessed children referred for concerns about progress with milestones or behaviour. I tested everything from motor skills, language, comprehension, attention and memory, to social and emotional behaviour. There were so many children. So many concerned parents – worried about autism, worried about attention deficit hyperactivity disorder, or worried that their child's IQ was lower than the 'norm'. I encountered many families in extreme adversity and saw children in very stressful situations. I am sure that we helped some of these children and their families through early-intervention and positive-parenting programs, occupational therapy, physiotherapy, social-support systems and medications when needed. But in some cases, where school failure was leading to antisocial behaviour, I was left almost certain the child was on a slippery slope with no easy fix. Once some situations are established they are hard, but not impossible, to break.

UNRAVELLING THE MYSTERIES OF THE BRAIN:

Our brain is the most complex, most amazing and most mysterious thing we have. We still really have no idea *exactly* how it works. But there have been significant advances in our understanding of the brain and its development, particularly in how early events, exposures and experiences can shape its neural networks.

Give or take, the mature adult brain is made up of around 100 billion neurons.[1] Neurons are the cells that process information throughout the nervous system, and form the intricate networks necessary for thoughts, feelings and actions. Each neuron can make 'synaptic' connections with more than 1,000 other neurons, and by adulthood it is estimated that we have more 60 trillion neuronal connections. Neurons have hundreds of short projections called 'dendrites', like the small branches of a tree, that receive electrochemical signals from countless other neurons. They then relay their own signals though one main output channel called an 'axon', which can extend across large distances to connect to other neurons. Axons are the data wires. Where the axon reaches the target region it forms synapses (connections) with other neurons sending electrochemical pulses to the target neurons.

This all happens at lightning speed. Where there are groups of neurons acting together in one region of the brain, their axons are bundled together for form fibre tracts that connect to other regions – like data cables. And just like data cables, axons are wrapped in insulting material. This white fatty substance, called 'myelin', increases the signal for high-speed data transmission. These parts of the brain, dense in myelinated axon cables, are referred to as 'white matter'. They generally sit below the 'grey matter' where most of the neuronal cell bodies are situated, on the outer surface of the brain.

It is in this thin surface layer of grey matter that most of our higher-level information processing arises – our neocortex. Our neocortex is linked to a series of critical information relay stations much deeper in the brain, called 'subcortical nuclei', also rich in neurons and also grey in colour. Together these are the largest and most important processing networks in the brain.

BUT IS IT NEUROLOGICAL OR PSYCHOLOGICAL?

When the brain is injured in a particular area, we see loss of function or abnormal function depending on the *region* affected, the *extent* of the injury and the *timing* of the injury. Localised defects, such as a stroke, have enabled accurate pinpointing of the neurons that determine particular functions. Different areas of the brain control speech, limb movements, and the detection of sensations in specific body parts. Damage to these areas produces predictable effects, with subtle individual differences. Some rare and curious defects lead to specific inabilities, such as being unable to recognise faces, or form new memories. There are also savants with uncanny mental abilities, photographic memories, calendar calculators, musical, artistic or mathematical genius. There are fascinating cases where these superhuman abilities appear *after* a brain injury! Many of these complex abilities work through networks of connections across multiple locations.

Some associations between anatomy and function have been recognised for a very long time. In times past, an obvious physical functional or sensory defect could be directly linked to physical defects in the brain at autopsy. These could be easily understood as 'neurological disorders'. Today, we can usually locate these kinds of focal defects with modern imaging technology while the patient is alive.

For other abnormalities of mental function, behaviour or emotion – such as schizophrenia, anxiety, depression or mania – few, if any, visible abnormalities were identifiable on autopsy. These 'psychiatric disorders' were harder to measure, define, and understand. Historically, 'neurological' and 'psychological' disorders have been defined and examined fairly separately, possibly because when these medical disciplines evolved, mental illnesses could not be explained by any obvious physical defect.

Modern imaging technology has now revealed subtle differences in brain structures in almost 50 per cent of people with conditions such as schizophrenia.[2] Functional differences can also be detected by patterns of blood flow and electrical activity, mostly in the frontal and temporal lobes, and some subcortical nuclei.[3] This suggests that there are also physical differences underlying 'mental' illnesses. There is growing evidence that these are the result of 'altered neural circuits' and the patterns of connectivity that occur during early development.[4] There is some overlap between 'neurology' and 'psychiatry', and one of the first steps a psychiatrist takes in making a diagnosis of mental illness is excluding another underlying neurological or other medical causes – ever aware that tumours, epilepsy and other conditions can produce psychiatric symptoms.

So this historical split between 'neurology' and 'psychiatry' remains, and it is reflected in two separate specialities of medicine. And it seems to work in practice. If we are depressed or anxious we look for a psychiatrist, but if we have epilepsy, weakness or numbness we find a neurologist. But these distinctions are still debatable, both in terms of the cause of each condition and in terms of the general understanding of brain and mind.

Many different conditions affect the brain, and in this chapter I focus on those that are most common, or most relevant to our background story of 'early programming'.

PREJUDICE AND THE DIAGNOSIS OF MADNESS:

Notions of possession, evil spirits or witchcraft were prevalent during the Middle Ages. During this period, those associated with these perceived evils – particularly women practising early forms of medicine against the edict of the Church – could be burned at the stake. These notions were gradually replaced with the awareness that 'madness' might be physical illnesses rather than a reflection of moral or spiritual corruption. But these illnesses remained ill-defined and poorly treated until the late nineteenth-century.[5]

History painfully shows that some of the 'behaviours' that we define as 'illness' are to some degree a social construct, determined by societal attitudes and belief systems. A modern example is homosexuality, which until quite recently was officially defined as a mental illness based on opinions deeply rooted in religious beliefs. Although declassified as a 'disease' by the American Psychiatric Association in 1973, some people still wrongly view homosexuality as a mental disorder.

Hungarian-born New York psychiatrist and academic Thomas Szasz, a well-known social critic, asserted that institutional psychiatry has been a tool of social control by medicine in modern society. In 1970, in his book *The Manufacture of Madness,* he likened institutional psychiatry to a modern version of the witch hunts.[6] While his views remain highly controversial, they prompted more public and critical examination of the role of psychiatry. There is still much concern that many behaviours, and even normal psychological processes, are still over-defined and over-medicalised.

There is an obvious continuum between mental health and mental illness – and these degrees sometimes make it hard to draw a line and give a label to someone. There is a range of normality and many of us have a number of traits that, if more extreme, can affect our relationships and our functioning in society, and increase our vulnerability to mental illnesses.

Our habitual reactions and views of the world can certainly evolve as we age, but the foundations of our personalities and our vulnerabilities are firmly laid early in our development. Patterns of 'connectivity' of our neural networks are determined by the complex early interplay between our genes and our environment while the brain is developing. This critically determines our individual capacity, our resilience and our later vulnerabilities.

SOME VERY COMMON PROBLEMS:

Ironically, although mental-health disorders are among the most common chronic illnesses in our society, they are also the most stigmatised. More than one in three of us will meet the diagnostic criteria for at least one mental illness at some point in our life, and this is true in most regions in the world.[7] Based on US data, anxiety disorders are generally the most common (29 per cent), followed by mood disorders (21 per cent), impulse-control disorders (25 per cent) and then substance-use disorders (15 per cent).[8-10] This pattern is seen in most parts of the world, although there are regional variations.[11]

The prevalence of personality disorders is less well studied, but appears to be around 13 to 15 per cent based on both European[12] and US surveys.[13] Psychotic disorders, such as schizophrenia, are far less common, with only 0.4 per cent of the population affected during their lifetime.[14]

When it comes to children, between 7 and 10 per cent of pre-school children are diagnosed with a psychiatric disorder or have significant emotional or behavioural problems.[15] Attention deficit hyperactivity disorder (ADHD) is the most common disorder, estimated to affect 6 to 7 per cent of children and adolescents using the DSM-IV criteria.[16]

Other important early-developmental disorders, such as autism spectrum disorder (ASD), are much less common, at around 1 per cent, but appear to be increasing. Part of the reason for an apparent striking twenty-five-fold increase is a change in the diagnostic classification. The new definition of autism has expanded to include previous diagnostic conditions such as Asperger's Disorder, Autistic Disorder, and 'Pervasive Developmental Disorder – Not Otherwise Specified' (PDD-NOS) into the single umbrella diagnosis of 'Autism Spectrum Disorder'. Even so, the broadening of the diagnostic boundaries cannot fully account for the increase in prevalence.[17]

And, finally, a word about dementia, as the most unlikely player in this story. There is some evidence that early-life exposures to toxins, variation in nutrition or maternal care, and other stressors could have latent effects that predispose to Alzheimer's disease.[18-20] In most regions dementia affects 5 to 7 per cent of those over sixty with almost 60 per cent of these living in low- or middle-income countries.[21] These numbers are projected to almost double every twenty years, although one UK study has shown a slight reduction in dementia rates.[22]

Although these conditions are extremely diverse in their pathology and symptoms, there is evidence that a range of factors (described following) that influence various aspects of the developing brain can predispose to both early- and late-onset disorders of the nervous system.

THE IMPORTANCE OF TIMING IN BRAIN DEVELOPMENT – DYNAMISM VERSUS DETERMINISM:

When it comes to the developing brain and nervous system, a basic understanding of *what* is happening and *when* can help us identify potentially more vulnerable periods. One of the best-known examples of this is the timing of neural-tube development and vulnerability to spina bifida.

During the third week of foetal development the 'neural tube' begins to form, as the first well-defined neural structure. This starts as two long ridges that rise up in parallel either side of the midline on around day 21 after conception. Over the next week these rise up and curve inwards to meet and fuse, forming the hollow 'neural tube'. The 'top' end of this tube will later become the brain and the 'tail' end will form the spine and spinal cord. Neural-tube defects such as spina bifida occur when the neural tube does not 'close up' completely at this crucial stage. Folate and other associated nutrients are important in this process, probably because of their role as 'methyl' donors in 'epigenetic' regulation (Chapter 11) of gene expression. This is the reason for ensuring adequate folate levels in women *very early* in pregnancy. Foods in most regions are now fortified to increase the likelihood of adequate levels in more women preconception. High doses of folate after this period do not prevent spina bifida and may increase the risk of other disorders, such as allergic diseases (Chapter 10).[23, 24] This needs further investigation, but highlights the importance of 'timing'.

Our genes depend on the environment to shape and influence the direction of the emerging neural networks (epigenetics; Chapter 11). *'Brains do not develop normally in the absence of critical genetic signaling and they do not develop normally in the absence of essential environmental input'.*[25] The external environment plays a role in shaping even the embryonic brain. As the foetal body develops, its own 'internal' environment adds another dimension to environmental cues. In fact, there is mounting evidence that the developing brain is influenced by internal metabolic and immune factors.[26] These also interact with a wide range of factors in the external environment to influence the progression of brain development.

Another critical element is the 'time sequence' – development must unfold sequentially with each element being dependent on preceding events. The patterns and order of expression in networks of genes are critical, and can be shaped by environmental cues.

New studies reveal that conditions such as autism might be linked to differences in the 'patterns' of these gene networks and how they are synchronised.[27] In other words, not necessarily a problem with the gene 'blueprints' but rather a problem with how and when they are read and put into action.

Over the past three decades there has been a radical shift from strongly 'deterministic models' to this more 'dynamic and interactive model'. The sequential expression of gene networks give rise to the progressive differentiation of neuronal structures and the progressive commitment to functional systems.

This is a far more exciting and intelligent way of looking at the brain than the very 'anatomical' perspective. Although we still have much to learn about the mysteries of the brain, some of the answers will lie in the differences in connections and patterns of connectivity created in the early development of neural circuitry.

DEVELOPING NEURAL NETWORKS – A WEB OF INTRIGUE:
'Network Theory' seems to be springing up in every aspect of our lives. Patterns of network connectivity are used to explain everything from how the internet works, how viruses like HIV or influenza spread, and how electricity grids work, to how thousands of crickets chirp and fish school in complete synchrony. This recent discovery – that there are universal laws that govern how all 'networks' work, no matter what they are – has enormous implications for almost every aspect of life. Already, in almost every field of medicine, these mathematical models are now being frantically used to link disease risk to *patterns* of gene expression in ways that we never could before. There is great excitement that this will bring better understanding of diseases and reveal new treatments and preventative measures.

The *dynamics* of how our neural networks are established, and the eventual patterns of connectivity, will determine some of our individual differences in function and capacity. You might expect the number of synaptic neural connections in our brain to increase progressively between infancy and early adulthood as we learn and experience. But no – it is almost virtually the opposite! The number of connections (synapses) in the infant brain is nearly *twice as great* as in the adult brain. Early 'synaptic exuberance' gives way to gradual 'pruning' across childhood and adolescence to reach the normal adult levels. This process of pruning is a critical part of refining the network, and will depend heavily on environmental cues.

There also appear to be a number of 'transient connections' that develop throughout the brain – but that are not seen in adults.[28] Although temporary, these connections appear important for the developmental process.

Perhaps the most utterly fascinating dimension to this, to me at least, is how quickly these connections change over just hours or even minutes. At a microscopic level, individual neurons form and retract connections over very short periods of time. This suggests that axons very rapidly sample the surrounding tissues as they seek their targets, and form and retract their synaptic connections according to the local environment.[29] This rapid, dynamic and evolving process guides each axon to its eventual target neuron.

There is speculation that pruning and cell death might be a way of correcting errors made in the process of connection or migration. 'Programmed cell death' does appear to be a necessary part of the brain developing, particularly in closing down the transient early connections that are no longer needed in the mature brain.

The 'input' signals that neurons receive, from both internal and external stimulation, play a vitally important role in refining, stabilising or eliminating our neural pathways. We refine and adapt our neural organisation according to our environment and context. Animal studies demonstrate just how extensive these effects can be.

Animals reared in more 'stimulating' environments with changing landmarks and many littermates have differences in both brain structure and function compared to those reared in standard laboratory cages – most notably in the 'connectivity' of their neural networks.[30] The animals that experience more complex environments during their development have much greater density of cortical synapses. This is matched with greater numbers of brain-support cells, and greater complexity of the blood vessels that supply the brain. Once these networks are established, they persist as the animal matures, even if they are later placed in more impoverished conditions.

'Deprivation' in early life has the opposite effect. Animals raised in a restricted and deprived environment show reduced complexity of neural networks and many other physical and functional abnormalities, which make them less resilient to stress in later life.[31]

The effects of more *specific* sensory deprivation also reveal the remarkable plasticity of the brain. For example, if visual input is only received from *one* eye the neuronal region processing signals from the 'active' eye expands and takes over the territory of the 'deprived' eye. This shows capacity for neural 'reorganisation' within sensory systems in the developing brain. This degree of restructuring is no longer possible once the brain is mature.

Another example could be seen when researchers surgically blocked signals from the ear in newborn ferrets. Remarkably, without any auditory input, the auditory cortex was taken over and restructured by visual pathways.[32] This was possible because the visual and auditory centres have transient connections, not present in adults, which had not yet been 'pruned' at this stage of early development. Sensory input from the eye stabilised these connections and allowed that part of the brain to assume a new function.

This all shows that processes of brain developed are 'designed' to anticipate experience, and are dependent on it for normal development.

The early postnatal period is particularly important for organisation of neural networks in the neocortex, although early patterning begins in the embryonic period. However, the nature of our experience remains important throughout childhood as the brain continues to develop. Even our more 'mature' nervous systems continue to require stimulation for optimal function.

LASTING EFFECTS OF EARLY 'EXPERIENCE' ON BRAIN DEVELOPMENT IN HUMANS:

We intuitively know how important a nurturing, loving and positive early environment is for both physical and emotional development, and for future well being. And we also intuitively know, and too often see, that emotionally deprived or neglected children are more likely to develop emotional problems, drug dependency and antisocial behaviours when they are older.

There is research to support this. With a warm and nurturing family we are more resistant to stress and stress-induced illness.[33] Social deprivation in early life can have lasting physical effects and functional effects on the developing brain.[34] Maternal depression is associated with decreased maternal sensitivity,[35] and a greater risk of later depression in her child.[36] In animals and humans, deprivation is associated with early changes in gene expression and function that persist into maturity, and increase the predisposition to both mental illness and physical illnesses.

Children who experience emotional or physical adversity are at considerably greater risk of mental illness, obesity, and NCDs such as diabetes and heart disease in later life. At the extreme end of the spectrum, physical or sexual abuse in childhood has well documented long-term effects on mental and physical health.[37]

Children who experience persistent emotional neglect, family conflict, and harsh or inconsistent discipline are more likely to have impaired physical growth[38] and intellectual development.[39] As adults they are also more likely to develop obesity,[40] depression, and anxiety disorders.[41] This risk increases to the same extent as if they had been more overtly physically abused.

Childhood trauma can change the grey-matter volume in certain areas of the brain. Young adults who *witnessed* domestic violence during childhood (between parents) show reduced grey-matter volume and thickness in their visual cortex. The researchers found that there may also be potentially more vulnerable periods between seven and thirteen years of age.[42]

Even low parental bonding scores, reflecting more aloof parent–child relationships, have been linked with anxiety and depression later in life. This has wider implications for long-term health. In the 1950s a cohort of healthy Harvard undergraduate students completed questionnaires about their perceived feelings of warmth and closeness with their parents. Researchers reassessed their health thirty-five years later, and found that the students who had rated their relationships with parents as cold and detached (when they were in college) were substantially more likely to have chronic disease, including coronary artery disease, hypertension, stomach ulcers, depression and alcoholism.[43]

Our parents are usually our most meaningful source of social support in early life. So it should be no surprise that our perception of parental love and care appears to have such important effects on our physical and mental health throughout life.

The 'social gradient' is a major factor to consider here. Families and parents who experience poverty or adversity are more likely to experience negative emotions, irritability, depression and anxiety, which impact on bonding, attachment and the quality of parental care.[44] Parents under these pressures are more prone to verbal threats, controlling attitudes, pushing or grabbing, emotional neglect, or overt physical abuse of the child. These more hostile and coercive dynamics are associated with greater likelihood of emotional and behavioural problems in adolescence. This highlights the importance of economic factors and social inequities in mental ill health.

THE BIOLOGY OF MATERNAL NURTURING AND NEURAL PROGRAMMING:

Differences in maternal care in animals will actually change gene expression in the brain of her offspring with lasting effects. A part of the brain called the hippocampus appears to be particularly important in this. And this seems to be consistent with what we see in humans.

Studies in rats show these effects in detail. Pups that are raised by a low-nurturing, ignoring mother become stressed and anxious as adults, but those raised by a high-nurturing mother (who spends time licking and grooming her pups) grow up to be calm.[45] And this is not due to an inherited effect, because the same effect is seen when the pups are swapped at birth. In other words, the pups born to a high-nurturing mother will become anxious and stressed if they are raised by a low-nurturing mother.

These differences in maternal behaviour act by changing gene expression in the stress control pathways of the pup's brain. The pups might *inherit* the same genes, but the *expression* of those genes is changed by the pup's experience. In other words, this is an *'epigenetic'* effect (Chapter 11). One of the main genes affected in these animals is the Glucocorticoid Receptor (GR) gene that normally helps shut down the stress response.[46] Licking and grooming in the first week of life increases the level of GR expression, and this is still higher when the pups reach adulthood.

When we are stressed or perceive danger our brain switches on a 'stress response' that activates the release of adrenalin and cortisol for our 'fight or flight' response. This is all orchestrated by a system known as the 'hypothalamic-pituitary-adrenal (HPA) axis', which hormonally links the centres of the brain (hypothalamus and pituitary gland) to the adrenal gland that sits just above the kidney. This allows us to deal with a threatening situation, but if cortisol levels remain high there can be an increased risk of disorders including heart disease, depression, and susceptibility to infection. We switch off the stress response by a 'homeostatic' feedback loop (a balancing process). Cortisol circulates to reach the brain where it binds to GRs. As the amount of cortisol binding increases, this signals the GR to turn off the stress response. Rats (and people) with *higher* levels of GR respond more quickly to cortisol. Essentially, they recover from stress more quickly. The higher level of GR in pups of high-nurturing mothers explains their more modest stress responses.

These epigenetic changes in gene expression actually involve hundreds of genes that show differences in methylation and other epigenetic controllers,[47] suggesting that maternal care has wide-ranging effects on the developing brain.

But all may not be lost! In keeping with the incredible plasticity of the brain, some of these changes can be reversed, experimentally at least. The same Canadian researchers injected a drug into the brains of the low-nurtured pups to artificially remove the methyl groups that had switched off their GR gene. The result – they were more relaxed and less anxious as adults, and acted as though they had been high-nurtured.[48] Injecting methyl donors (which switch off genes) had the opposite effect. It is not known how this would translate to humans.

This goes to the heart of the *reason* that the maternal behaviour has any effect on the pup at all. By its very nature our epigenetic programs allow for flexibility in gene expression, so that we can *adapt* to our environment. If the pups have a 'harsh' early experience it is adaptive to *anticipate* a harsher environment later. In a dangerous environment the more anxious and guarded behaviour of the low-nurtured rat is an *advantage*. But in an abundant, relaxed environment, relaxed animals thrive but anxious animals do not do as well. The mother's behaviour reflects *her* experience of the environment and helps her offspring predict and prepare for the environmental conditions they will ultimately face once they become independent.

This is nature's remarkably clever way of *transmitting* the *experience* of the mother to the next generation, to improve resilience according to the environmental conditions. It is also possible, and likely, that other early-life exposures such as nutrition and toxins (below) also have epigenetic effects on the brain to alter the risk of neuropsychiatric disorders.

EVIDENCE OF EPIGENETIC VARIATION IN HUMAN BRAINS?

During the prenatal and early postnatal periods the epigenetic machinery in the brain (which determines which genes are turned on and off, and when) appears to be particularly sensitive to both environmental and genetic factors that can alter the level of DNA methylation.[49] For example, abnormalities of methylation can reduce the dendritic branches of neurons, and once these structures are mature, large-scale structural remodelling does not appear to be possible.[50] Early abnormalities in organisation can lead to both early-onset neurodevelopmental disorders such as autism,[51] or predispose to later-onset disorders such as depression or schizophrenia.[52]

So how do the epigenetic changes seen in anxious or depressed rats relate to humans, who are much harder to study? In studying autopsy specimens, it has been noted that there is lower expression of glucocorticoid receptor (GR) in adults with major mental illness including depression, bipolar disorder, schizophrenia.[53] Differences in the extent and anatomical distribution, which could reflect the variations in clinical patterns and severity seen in these disorders, are consistent with increased activation in stress-response pathways that have been seen in psychotic and severe forms of depression. These differences in gene expression are suggestive of epigenetic effects, but it is difficult to infer when these occurred.

Suicide victims studied at autopsy had lower levels of GR gene expression in the brain (hippocampus), compared to control subjects of the same age.[54] But notably, these epigenetic differences were only seen in suicide victims with a history of childhood abuse. This seems to be consistent with the epigenetic effects of early parental care in rats on GR expression, an important determinant of later stress response.

Also in humans, epigenetic differences in GR can be detected at birth. One study found that maternal mood in pregnancy was associated with the epigenetic regulation (methylation) of GR in cord blood. Although it was not possible to measure levels directly in the newborn brain, it is notable that maternal depressive symptoms associated with higher GR methylation (lower gene expression) increased cortisol stress response in infants when they were three months of age.[55] This indicates that the programming of the human *foetal* stress response is also sensitive to maternal mood.

Among the many genes expressed differently in the brains of people with mental illnesses, brain-derived neurotrophic factor (BDNF) has been of particular interest. This is one of the most active growth factors regulating neuron growth and differentiation. People suffering from depression show decreased BDNF levels in the hippocampus at postmortem and this is consistent with a smaller hippocampus size seen with depression.[56] It also fits with the reduced levels of BDNF seen in chronically stressed animals.[57] Antidepressant treatment increases BDNF levels. Similar profiles have been seen with other growth factors,[58] and this all suggests that these have an important role in the regulation of mood. Studies in mice have shown that prenatal exposure to low levels of environmental toxins such as methylmercury decreases BDNF expression and predisposes mice to depression.[59] This shows that early exposures can induce lasting epigenetic suppression of brain-growth factors in the hippocampus, with likely implications for humans.

Schizophrenia does not typically emerge until adolescence or adulthood, but its associated structural abnormalities begin early in brain development.[60] Schizophrenia patients show increased methylation (and decreased expression) of the 'Reelin' gene, which is important for migration and neuron 'connectivity'. This is one of many genes that may be implicated.[61] Environmental factors in the prenatal or perinatal period implicated in schizophrenia include maternal malnutrition,[62, 63] infections in the second trimester,[64] exposure to inflammation for other reasons, and peri-natal injury. It is possible that these events, which are not specific to schizophrenia, may subtly alter the epigenetic regulation of gene expression and lead to neurodevelopmental abnormalities. This is not yet clear. What is interesting is that many of the genes associated with schizophrenia are important for neuronal growth, migration, and synapse formation,[65] and many are also implicated in autism and other neurodevelopmental disorders.[66] The most perplexing question remains why similar gene variants or early environmental insults produce early effects in some individuals (autism) but delayed latent effects (schizophrenia) in others.

In autism spectrum disorder (ASD), genetic inheritance is a major factor, but there is still strong interest in how prenatal environmental exposures may have predisposing epigenetic effects. There are many documented alterations in the anatomy and the connectivity of neural networks in children with ASD. There have now been a number of studies in humans that show that differences in overall gene methylation patterns in both parents of autistic children and in children with ASD. There are also studies showing increased methylation (and reduced gene expression) of specific genes implicated in autism.[67]

Sometimes, only *one* identical twin develops ASD and the other doesn't. The fact that this happens indicates that ASD is not *all* due to the inherited genes (which are identical). Such studies have shown increased methylation in certain genes of the twin with ASD compared with the unaffected twin.[68] This suggests that early epigenetic effects play an important role in ASDs, and there is speculation that both nutrition and metabolic factors could increase susceptibility to ASD through epigenetic effects. This is still being investigated.

Ultimately it is the *combination* of early environmental *adversity* and the *maternal reaction* to this that determines the initial programming of the offspring's stress response. If parents are resilient to environmental adversity, poverty and economic stress, their children are less likely to show adverse effects in emotional and cognitive development.[69] And the offspring's *later* environment is also important. This has been described as '*the three-hit concept of vulnerability and resilience*'; hit-1: genetic predisposition, hit-2: early-life environment, and hit-3: later-life environment.[70] It also returns us to the concept of 'match or mismatch' – the degree to which the conditions that the offspring are *programmed* to anticipate match what they *actually* encounter (as we saw with obesity in Chapter 4). When there is *mismatch* between early- and later-life experience, vulnerability to stress, anxiety and depression may be greater.

AN INTIMATE RELATIONSHIP BETWEEN OUR IMMUNE SYSTEM, OUR BRAIN AND OUR BEHAVIOUR:
Once again, we return to the immune system, which has an intimate relationship with the brain and with the 'HPA' stress-hormone axis. Although we have been focusing on neurons so far, the brain is also packed full of other cells. Many of these are now understood to be immune cells, and almost all cells in the brain, even neurons, produce 'cytokine' immune signals. It has been

recently discovered that these cytokine signals play an essential and unexpected role in learning and memory. This gives us a completely different perspective on the immune system.

The immune system is not just a defence system. It reaches into almost every part of the body and is constantly sensing the internal and external environment. And it is a core part of the neuro-endocrine processes that operate to achieve balance throughout all body systems.

But it is a relationship that works both ways. The effects of the *brain* in the immune system have been recognised for a long time — stress and anxiety make us vulnerable to infections and viruses, even if we don't fully understand how. But the idea that the *immune system* has major effects on the brain gives us new perspectives on how neuropsychiatric disorders develop, and how they might be treated or prevented.

The 'microglial' cells in the brain are immune cells, and closely resemble the 'macrophage' scavenger cells that are seen in other parts of the body (such as those seen in the blood vessels in atherosclerosis, Chapter 6). This network of cells is present in the nervous system from the early embryonic period, and plays a critical role in neuronal 'pruning' and elimination of suboptimal or redundant synapses. Defects in this pruning process are linked with autism (ASD) and schizophrenia.[71, 72]

Immune cells produce a variety of cytokines signals, ranging from inflammatory cytokines which include interleukin-6 and interleukin-1 (Chapters 6 and 10), to anti-inflammatory cytokines and growth factors. These are produced by most cells in the brain and modulate basic functions such as sleep, memory and metabolism.[73]

But cytokines change our behaviour. For example, when we are sick we become less active, less social, sleep more, eat less, and drink less. This 'sickness behaviour' is caused by the cytokine immune responses. The same 'sickness behaviour' can be induced artificially by infusing cytokines directly even in the absence of illness.[74]

These 'sickness behaviours' are strikingly similar to the behaviours and physical symptoms of depression and other neuro-psychiatric disorders. It may be no coincidence that immune abnormalities are also a feature of major depression, and a range of other disorders including schizophrenia, autism, Rett syndrome, anxiety and stress disorders. Increased levels of circulating inflammatory markers, cytokines, are also a feature of post-traumatic stress disorder (PTSD), which also shows a range of immune abnormalities and epigenetic changes in methylation of immune genes.[75]

Because of these associations, and the closely synchronised development of the immune and neuronal systems of the brain, it has been proposed that the immune system may mediate some of the early-developmental abnormalities seen in a range of neuropsychiatric disorders.[76] There are a number of ways that proinflammatory cytokines can affect mood and behaviour, such as activating the 'HPA' stress axis, or altering metabolism of the neurotransmitters in the brain. They may even have more direct actions on neural function (below) that have more significant implications for the relationship between the developing brain and the developing immune system.

EARLY-LIFE INFLAMMATION AND INFECTION AS A RISK FACTOR FOR NEUROPSYCHIATRIC DISORDERS:

Recent studies show a strong relationship between early-life infection and later-life schizophrenia, supporting a long-held theory. High levels of pro-inflammatory cytokines (interleukin-6 and interleukin-1) in either the mother or the foetus have been associated with abnormal foetal brain development and neuro-developmental disorders.[77] In particular, maternal influenza, which increases immune activation and cytokine synthesis, has been linked to a higher risk of schizophrenia in her offspring.[78]

In rat studies, infections introduced in pregnancy and early postnatal, as expected, cause *immediate* fever responses and activate the HPA stress axis (increased cortisol production). But there are also *long-term* changes in both immune and stress responses.[79, 80] Adult animals that were exposed to infection in utero show *behavioural* abnormalities consistent with symptoms seen in humans with schizophrenia and autism, including abnormal startle responses, decreases in exploratory behaviour, and decreases in social interactions. There is also an association with learning impairments and reduced ability to ignore irrelevant information in the environment, a cardinal feature of humans with schizophrenia. Many other bacterial products have been tested in the same way, including the 'endotoxin' product (Chapter 4). The effects vary. But overall, a consistent picture emerges – that infection when the brain is developing has long-term effects on adult immune function, stress responses and behavioural patterns.[81]

In severe and profound infections, we see clear irreversible damage on the developing brain. But in other cases the effects are much more subtle and delayed, and the association much less obvious. In this latter scenario there may be no obvious effects on the brain, and any future effects on behaviour are more dependent on the adult environment and how the immune/HPA stress systems might have been reprogrammed to respond to it.

Just to make that a little clearer, we can turn to see how this works in animals. Not *all* rats given the same 'infection' as newborns will develop learning or memory problems as adults. Only rats that have the *combination* of the neonatal infection and a later immune challenge that induces inflammatory responses develop the memory defects. This effect can be prevented by blocking the interleukin-1 inflammatory response showing that, at least in this animal model, the adverse effect on memory and learning was caused by the exaggerated inflammatory response.[82] Although artificial and simplified, this fits with the 'two-hit hypothesis' of schizophrenia: an underlying vulnerability (established early in life) *plus* a later-life precipitating event (such as stress or infection) usually in early adulthood, is required for the appearance of the illness.[83]

These observations have also led to greater awareness of the remarkable role that immune cytokines, such as interleukin-1, play in *normal* brain function, particularly in the hippocampus region of the brain. The very *process of learning itself* will release interleukin-1 in the hippocampus.[84] The hippocampus plays a central role in regulating stress responses, but it is also critical for learning, memory, emotion, and other survival behaviours. Either too much or too little interleukin-1 can impair learning and memory. Simply injecting interleukin-1 into the hippocampus induces memory impairments in rats, and changes measures of synaptic function in the hippocampus.

In humans we can see exaggerated interleukin-1 responses in conditions associated with cognitive impairment.[85] A good example is Alzheimer's disease – a recognised inflammatory condition with higher levels of interleukin-1.[86] This might explain why lifestyle factors that promote inflammation, such as a high-fat diet[87] and physical inactivity[88] are also risk factors for Alzheimer's disease as well as many other chronic inflammatory NCDs. Abnormal interleukin-1 responses in depression and anxiety disorders also fit with the associated impaired memory, learning and 'sickness behaviours'.

For these reasons, it is important that interleukin-1 is very tightly regulated during the course of the immune response within the normal brain. The microglia cells are the main source of interleukin-1 produced in the hippocampus, and are most affected when early-life infections 'shift' the responses of the adult rat towards a more exaggerated interleukin-1 response and memory defects.

To summarise – microglia play a critical role in learning, by regulating interleukin-1 and the synaptic mechanisms underlying memory in the hippocampus. Any factors that lead to excessive or exaggerated 'inflammatory' microglia responses can impair learning, memory and behaviour. And early-life events can influence our later cognitive function through the programming of microglial function. But, of course, our ongoing exposures and immune experiences are also important in shaping our patterns of health and disease susceptibility.

. . .

Although there is still some inherent flexibility in the immune system, many environmental factors (such as nutrients, pollutants, and microbial exposures) may have early epigenetic effects on immune gene expression, and immune diseases such as allergy appear in the first months of life, suggesting a quite early 'point of no return' (Chapter 10).

Once again, it is likely that if there is 'mismatch' between what we anticipate and what we encounter, there will be a greater risk of inappropriate immune responses that can lead to disease. However, the immune system is complex – immune responses are varied and tailored to exposures via experience – and we are yet to understand exactly how the simple 'mismatch' theory applies here.

Like the brain, the immune system also carries millions of memories – it remembers everything we have seen (Chapter 10). Things go wrong when memory cells (T cells and B cells) make the wrong decision and react to harmless things. If the scavenger macrophages (which are very similar to the microglial cells in the brain) give 'danger' signals to the T cells they are more likely to react inappropriately.

With the very intimate relationship between the brain and the immune system, it is logical to suspect that there might be a connection between the rising predisposition to immune disease and the predisposition to mental-health disorders.

The immune system is highly susceptible to the environment in early life, and we must not overlook the possibility that this early immune dysregulation could also be affecting the developing brain, which is also highly vulnerable during this same period. Future research should investigate this, as promoting early immune health could be an avenue for the prevention of mental illness.

INFLAMMATION AND THE WIDER ENVIRONMENTAL DETERMINANTS OF MENTAL HEALTH:

'Infection' is an obvious source of immune activation or inflammation in early life. But nutritional patterns, environmental toxins, physical activity, and maternal-health factors including obesity, gestational diabetes, preeclampsia (high blood pressure) in pregnancy, and a range of other factors and stressors, can influence both growth and inflammation.[89] As the immune cells of the brain, microglia are very sensitive to any inflammatory triggers in the systemic circulation. They are able to detect the same broad range of 'danger signals' as other immune cells, including microbial signals and *non*-microbial danger signals. These can be in the external environment or 'alarmins' that are released from damaged tissues and inflammation within the body. This all means that the microglia may be potentially re-programmed by a range of other inflammatory events early in life, to modify developing brain function.

In this light, we need to reflect on the implications of the increasing tendency for chronic low-grade inflammation associated with modern-lifestyle changes. Many lifestyle factors (discussed in Chapter 6) promote chronic inflammation, and these appear to contribute to significantly higher baseline levels of inflammation measured by levels of C-reactive protein (CRP). The implications for mental health are less clear than for other inflammatory diseases such as cardiovascular disease and diabetes. Although preliminary, there are new studies linking higher maternal CRP levels in pregnancy with an increased risk of autism in offspring.[90] This will be an important future direction of research. Even though animal models do not mimic this situation, they do reveal that adult microglia function is modified by early environmental factors.

The relationship between metabolic health and immune health is a recurring theme, and it is also clear that metabolic health is of major importance to the nervous system. One of the central metabolic goals is a steady, stable nutrient supply to the brain and vital organs. Many metabolic diseases impact on the brain. And in the newborn period, metabolic derangements such as extreme jaundice or very low blood-sugar levels (Chapter 5) can produce permanent irreversible brain damage if untreated.

The effects of milder metabolic derangements on the brain are less well studied. The chronic effects of obesity on the immune system, for example, are well recognised, with increased levels of inflammatory cytokines (interleukins), but it is not clear whether higher rates of dementia are the result of direct inflammatory effects on the brain, or secondary effects of other NCDs such as cardiovascular disease and diabetes. Other interrelated factors, such as low physical activity, are also risk factors for dementia. There are likely to be more complex effects of the metabolic–immune interplay on the brain than we have previously recognised. Early-life factors such as heavy metals, high-fat diets and other inflammatory factors may have 'latent' effects on gene regulation.[91] These long-term changes increase the risk of dementia and more rapid neurocognitive decline much later in life.

There are risk factors implicated in *both* immune and brain development (risk factors for inflammation alone are discussed elsewhere). So far, these effects have only been studied separately and we still don't know the relationship between the effects, or if early interventions would benefit both.

The first is nutrition. Malnutrition ('bad' nutrition; both under- and over-nutrition) is becoming more common. Foetal growth restriction, even if mild, is a risk factor for neuro-cognitive difficulties and lower school achievement.[92] Extreme under-nutrition has well known adverse effects on immune function and brain development.

The Dutch Hunger Winter at the end of World War II (Chapter 4) allowed researchers to look at the effects of starvation at specific times of pregnancy on many aspects of neuropsychiatric health.[93] The offspring of women who experienced famine in the first two months of pregnancy or a month before conception were at much greater risk of spina bifida (neural-tube defects), as well as personality disorders and schizophrenia in adulthood. Clearly, these effects could have been mediated by many different nutrients. But the coincident increase in both neural-tube defects and other neurodevelopmental disorders raised the question of whether folic-acid supplements (important in early development to prevent against spina bifida) could have broader preventive effects, particularly as they have significant epigenetic effects.

Many other nutrients are believed to be of importance in immune development and/or brain development including zinc, polyunsaturated fatty acids, fat-soluble vitamins (e.g. vitamins A, D, E), choline, and B vitamins,[94] but their role in preventing associated diseases is not yet established.

The second risk factor worth noting here is environmental toxins: airborne pollutants and chemicals that contaminate our food and water. A number have been implicated for their neurotoxicity, immune toxicity, or both. The damaging effects of heavy metals, such as lead, on the developing brain is a good example – and also shows how a simple public-health measure (removing lead from water pipes and children's toys) can prevent early brain damage.

Exposure to even moderate air pollution during late pregnancy can influence cord-blood cytokine secretion in healthy neonates.[95] It can also induce epigenetic changes in immune cells to promote inflammation and allergic response.[96] These inflammatory changes have implications for the brain, and preliminary studies in animals do show how prenatal air-pollution exposure can induce inflammation, activating microglia in the brain.[97] Our knowledge of these interactions is still fairly rudimentary, and is important to pursue further.

For many years there has also been evidence that early exposure to chemicals such as polychlorinated biphenyls (PCBs) is associated with neuro-developmental abnormalities,[98] overt brain damage[99] and immune effects. These toxins have also been linked to specific conditions such as autism spectrum disorders.[100] More than half the chemicals we are exposed to are untested for toxicity. In particular, the effects on developing organs, such as the brain, during the most vulnerable periods of development need to be understood.

Maternal exposure to the by-products of disinfectants used in public-water supplies has been linked to a range of adverse outcomes, including pregnancy loss, prematurity and low birth weight, as well as birth defects, but this is far from conclusive.[101] Even so, disinfectants might also be having other indirect effects – by changing the pattern of 'normal bacteria' in our environment. Which leads us to another recurring theme.

IS THERE A GUT–BRAIN AXIS?

Although the idea that the gut and the brain might be linked is not new, the biology behind this,[102] and central role of the microbiota,[103] is only just emerging. Apart from the brain, the gut is the other site where immune networks and neural networks interact in vast numbers and complexity. The large neural networks of the 'enteric nervous system' provide a direct 'neural' link between the gut and the central nervous system. This allows two-way communication to control gut function and relay sensory signals from the gut environment.[104] There are also 'hormonal connections' between the brain and the gut, in particular the parts of the brain that control appetite and 'stress' responses. The immune system provides another link between the gut and the brain. Immune development is almost completely dependent on gut bacteria – at the very least, this has indirect implications for the brain.

There are now many studies suggesting bacteria could influence brain function and behaviour, even if we don't yet fully understand how.[105] Mice raised in 'germ-free' conditions show the now-well-recognised failure of normal immune development.[106] But they also show differences in gene expression in the brain and abnormal stress responses. In fact, they have reduced expression of the neuronal growth factor BDNF in the cortex and hippocampus,[107] consistent with patterns seen in depression and anxiety. And in line with this, they also have heightened cortisol stress responses. This shows that without normal bacteria, 'germ-free' animals will develop potentially maladaptive changes in brain function and in their endocrine regulation of the stress response through the HPA axis.

What is even more fascinating is that these effects can all be reversed by either giving 'friendly' probiotic bacteria, or by a much less savoury 'faecal transplant' from a normally colonised mouse.[108] But the real 'bottom' line is that timing is critical – reversal is only possible at an *early* stage of development. This is true of both the stress responses and the immune responses, which remain abnormal once they are 'committed' – though even in mature animals, disrupting the established normal gut flora with antibiotics can result in changes in BDNF in the brain.[109]

There are now many studies supporting the gut–brain link, and probiotic bacteria, in addition to other treatments, is being considered as a potentially useful adjunctive therapy in patients with depression and other neuropsychiatric conditions.[110]

Using a double-blinded trial, a Californian team recently showed that healthy women who took probiotics twice a day showed altered brain function after four weeks, both in resting brain activity and in response to an 'emotional-attention task' compared with women who had a placebo.[111] The researchers are confident that this could mean that probiotic strains in yoghurt could have wider health benefits than previously thought, such as relieving anxiety, stress, and other mood symptoms over time. There is also some evidence that probiotics change pain sensitivity, which may be one reason for the effectiveness in treating irritable bowel syndrome.

For developmental conditions such as autism, links between the gut, the immune system and the brain have long been suspected. A high proportion of children with autism spectrum disorders (ASD) also have gastrointestinal symptoms and immune abnormalities.[112] Higher levels of inflammatory cytokines have been detected in the circulation of these children,[113] and it has been proposed that this may be the result of increased intestinal permeability[114] and activation of immune responses, possibly as a result of abnormal gut flora. Clinical trials are now needed in this area. And this is likely to only offer part of the answer to a complex problem.

There is little doubt now that colonisation of the gut plays a major role in the post-natal development and maturation of the immune and endocrine systems. This all has implications for the developing brain.

An important goal is to minimise unnecessary or inappropriate inflammation (Chapter 10). If the microglial behave like other immune cells, and there is every reason to believe they do, we might expect them also to be more activated with chronic inflammatory states. In general, we might argue that, at least in general, 'what is good for the immune system' is also likely to be 'good for the brain'. That factors that promote the 'optimal' immune development are likely to also be of benefit in reducing the burden of neuropsychiatric disease.

SPENDING MORE TIME IN THE GARDEN:
Although we intuitively feel healthier in nature, there is now a scientific basis that this may influence our physical and mental well-being.[115-118]

I first learned about this during a trip to Japan in 2003, where I met Dr Alan C. Logan, one of the leading proponents of the importance of our relationship with nature.[119] For many years Japanese researchers have been investigating the health benefits of *shinrin-yoku* (literally translates as 'forest-air bathing') or walking in a forest environment. Studies have since shown that, compared to city walking, forest walking is associated with favourable physiological effects including lower blood pressure, lower heart rates, lower levels of stress hormones (cortisol) and better mental outlook and vitality.[120] They even saw differences in the electrical activity of the brain.[121, 122]

This is consistent with studies in large cities of Europe and North America, showing that more 'green space' in proximity to people's living environment is associated with better health and reduced mortality. Spatial planning policy and preserving green spaces has enormous economic implications through savings in health costs.

Proximity to parks may encourage natural light exposure (with the benefits of vitamin D and moderate UV) and physical activity, which also has beneficial health effects. But there appears to be more to it than that.

Even just *looking* at photographs of natural scenes induces difference patterns of brain activity.[123] This can be seen using sophisticated imaging methods (functional MRI studies) to study brain activity. When healthy subjects looked at scenes of rich vegetation there was a relative increase in activity of part of the areas associated with emotional stability (such as anterior cingulate) and feelings of love (the insula). On the other hand, viewing the urban environment increased activity in areas of the brain associated with fear, threat or danger (such as the amygdala), which are also active in people with anxiety and depression. These responses occurred very rapidly, suggesting that they are instinctual reactions and reflect our primitive links to the natural environment.

This also translates into the health setting. Patients placed in hospital rooms with a 'tree view' show quicker recovery than those in rooms with a view of a brick wall.[124] They also show fewer complications and less need for pain medications. In healthy people, being around plants in the home and workplace can improve concentration, memory and attention span. Memory performance is better after a nature walk compared with a city walk, and simple and brief interactions with nature can produce a calming influence and marked increases in cognitive control.[125]

This has important implications for children in terms of both behaviour and school performance. Both plants in the classroom[126] and a view of natural landscapes[127] have been associated with improved performance and test results in normal children. For children with attention deficit hyperactivity disorder (ADHD), time in nature also has beneficial effects on concentration and performance. A 20-minute guided walk in the park is sufficient to increase attention performance compared to when children spent the same time walking in a concrete urban landscape.[128] This suggests that 'doses of nature' are a safe, inexpensive, accessible tool in managing ADHD.

WHAT MIGHT THE FUTURE HOLD?

It is important to think more holistically about the *many* factors that contribute to the development and decline of our cognitive abilities. Everything we are discovering particularly points to the need to take a more unified approach that simultaneously examines the interactions between the developing *immune* system, the developing *stress* responses (adrenal axis), developing *metabolic* responses and the developing *nervous* system in a much more integrated way.

It also means that we need to simultaneously consider the health and development of all organ systems when we consider our brain, because they are all interconnected.

We have come so far, and there is still so much to learn. A more integrated approach will ideally consider all aspects of development, ranging from physical and emotional experiences through to metabolic and immune experiences. This predictably brings us back to the same story – physical activity, reducing toxic exposures, and a healthy balanced diet which we hope will also afford us more healthy gut bacteria in the process.

8

Bones muscles and joints

Our ability to move freely, without pain, is a crucial aspect of our general health, our fitness, our quality of life and even our life expectancy. Too often this is something we take for granted until something goes wrong. And its general importance to so many other aspects of our health is often underestimated.

Our bodies are designed to move, and even after relatively short periods of bed-rest or inactivity we lose muscle strength, cardiovascular fitness, and physical endurance. More prolonged limitation of physical activity, such as occurs with arthritis, will increase the risk of obesity, diabetes, high cholesterol and the vulnerability to heart disease. These effects are likely to be due to a combination of adverse changes in metabolic balance and the loss of the inflammatory effects of exercise. Depression is also more common, probably as a result of both the physiological effects of inactivity and psychological impact of pain, frustration, stress and fear for the future. Immobility also leads to loss in bone density, adding to the risk of facture and further movement limitation. At the extreme we see many other consequences, such as poor circulation, fluid accumulation, deep vein thrombosis (DVT), skin ulcers and pressure sores, constipation, acid reflux, joint contractures and fractures, and even death from embolism or pneumonia.

Too many times as a junior doctor I saw hip fractures lead to bed confinement, then to a rapid decline in health in previously active older women. Until the fracture, many did not even know that they had osteoporosis, or thinning of their bones. This silent condition has its origins much earlier in life when our peak bone mass and growth patterns are set.

The growing burden of musculoskeletal disease:
Musculoskeletal disorders affect the bones, joints, ligaments and muscles that form our 'locomotor' system and our body frame. Some are evident very early in life as the result of genetic disorders or congenital malformations. Some, such as rheumatoid arthritis, are due to inflammatory immune disorders and can appear at any age. And others are more linked to ageing and degenerative processes, including osteoporosis and osteoarthritis (OA). They share common elements: altered mobility and function, pain, and the secondary social and health consequences that these can bring, depending on the nature and degree of the disability.

The increasing burden of these diseases is largely linked to our ageing population, with degenerative conditions such as OA and osteoporosis the most common musculoskeletal disorders. Most of the OA disability burden is due to arthritis in the hips and knees, and OA is the primary cause for 90 per cent of the increase in hip- and knee-joint replacement operations worldwide. Large-scale population studies in countries such as the USA reveal that 21 per cent of adults over the age of eighteen report a doctor diagnosis of some form of arthritis, and 50 per cent of people over the age of sixty-five suffer from arthritis.[1] Most forms of arthritis are more common in women. In the developing world, populations are also ageing. Because of their sheer population size, China and India already have the greatest numbers of people over the age of sixty-five. The prevalence of knee pain in many Asian countries is already comparable to regions of the developed world.[2]

As well as the obvious personal cost, the annual financial cost of arthritis in the USA has been estimated at over $100 billion. More than 50 per cent of this is due to lost earnings and indirect costs. Arthritis sufferers also become less physically active, contributing to a higher risk of obesity, depression, heart disease and other NCDs, and amplifying the long-term health burden.

Osteoporosis occurrence also increases with age, as bone mass and bone density decline. Resulting fractures occur most commonly in the wrist, spine (vertebrae) or the hip. It has been estimated that 50 per cent of women and 20 per cent of men aged fifty will have an osteoporosis-related fracture in their remaining lifetime. After suffering a fractured hip, 50 per cent of people will be unable to walk without help and 10 to 20 per cent will die within six months.[3]

The global burden of osteoporosis is also set to increase markedly due to ageing populations. In 1990, there were an estimated 1.26 million hip fractures globally and this is projected to increase to between 6.26 million and 21.3 million by 2050, with catastrophic effects.[4, 5] In 1990, 26 per cent of all hip fractures occurred in Asia, and it is anticipated this could rise to 45 per cent by 2050. The impact will be more devastating in developing regions where there are fewer resources to cope with it. Better prevention strategies are clearly needed, with much earlier interventions to curtail or prevent osteoporosis.

Early interventions will help improve the strength and the resilience of our bones, and *ongoing* interventions will help reduce bone loss and joint damage as we age.

BONE STRENGTH IS DETERMINED EARLY IN LIFE:

Factors in foetal life, childhood and adolescence determine our maximum bone mineral density (BMD) and our 'peak bone mass' as young adults.[6] Additional factors in adulthood are then important for determining the rate of bone loss.[7] Anything that results in a lower BMD and peak bone mass, or a higher rate of bone-mass attrition, will set the scene for osteoporosis, bone fragility and fractures.

In this equation, our peak BMD, set early in life, appears to be even more important than either age-related bone loss or the age of menopause.[8]

Bone mineral density is typically measured by 'dual-energy X-ray absorptiometry' (also known as 'DXA'). DXA scans track the progression of osteoporosis in adults. It is also sometimes used to measure skeletal maturity and body-fat composition in children and to evaluate rare paediatric bone disorders. DXA has also allowed more detailed study of bone development and tracking of BMD and bone mass with age.

For girls, the peak period of bone mass accumulation is between the ages of eleven and fourteen, with over 95 per cent of peak bone mass acquired by the late teens. After puberty, boys acquire greater bone mass than girls. There may be small gains until around thirty years of age, when bones reach their maximum strength and density. After menopause most women experience relatively rapid initial bone loss, which then slows but continues throughout the postmenopausal years when the rate of osteoporosis increases markedly. For men, there is also age-related bone loss, but osteoporosis is less common because of their higher peak bone mass. Although there are genetic and racial differences in the risk of osteoporosis, a substantial component of the risk is determined by modifiable environmental factors.

Growth trajectories in early childhood are strong predictors of adult height, weight and peak bone mass. A lower rate of growth in childhood is associated with a significantly greater chance of hip fracture in later adulthood.[9] Our weight in early infancy also predicts our bone mass sixty-five to seventy-five years later.[10, 11] Long-term studies have shown that lower body weight during the first year increases the risk of osteoporosis fractures much later in life.[12] There is also evidence that reduced growth during foetal life is associated with changes in the structure of the hip bone that reduce the mechanical strength and risk of fractures in later life.[13] This has focused attention on maternal factors that influence foetal growth in general and, in particular, bone development.

FACTORS THAT INFLUENCE FOETAL BONE DEVELOPMENT:
The foetal skeleton begins as a cartilage model that can been seen by five weeks of gestation, formed by the precursor cells and stem cells that will, in turn, ultimately form bone. Growth hormone, insulin-like growth-factor-1 (IGF-1), vitamin D and vascular growth factors are all important for these processes. Once the cartilage model is formed, blood vessels grow into the structures, which are gradually invaded by 'osteoblasts' – the bone-forming cells. These lay down more matrix, which is then progressively mineralised. Calcium from the mother across the placenta is important for this process. Mineralisation of the organic bone matrix increases exponentially between twenty-four and thirty-seven weeks of gestation, during which calcium and phosphate salts are deposited.

A range of factors can interfere with these processes, and many of these have been linked to increased risk of late-life osteoporosis and fractures. Maternal diabetes mellitus, preeclampsia and hypertension, renal failure, malnutrition and low vitamin D status all have a deleterious effect on foetal skeletal health.[14] Anything that compromises placental function can interfere with calcium and phosphate transport. Babies with intrauterine growth restriction and placental insufficiency may also have reduced mineralisation ('osteopaenia'). Demineralisation is also seen in infants born to mothers with infection or inflammation ('chorioamniositis') of the placenta.

Vitamin D and calcium are both important for mineralisation in bone development, and vitamin D is also important for differentiation of cartilage and bone-forming cells. Lower maternal (and cord-blood) levels of both vitamin D and calcium are associated with reduced bone mass in later childhood. Vitamin D insufficiency is common. Lower vitamin D levels in mothers during late pregnancy have been associated with reduced bone mineral content in children at age nine.[15] Differences in vitamin D and sunlight exposure may also explain the lower bone mineral content in winter-born versus summer-born infants[16] and the different fracture rates decades later.[17] It has been proposed that vitamin D supplementation of pregnant women, especially during winter months, could have long-term benefits and reduce the future risk of osteoporotic fracture in their offspring.

The role of vitamin D supplementation in pregnancy is at the forefront of many discussions around how to provide optimal conditions for early programming and disease prevention. At present this is a very murky area. There is still much debate about what the optimal, normal ('sufficient') levels of vitamin D actually are. A high proportion of women fall below some of the arbitrary thresholds, and it is not clear what the health implications might be, if any, for those that only have mild 'insufficiencies' (as opposed to clearer 'deficiencies'). There are concerns that 'routine' supplementation of *all* women could lead to excessive levels and potential ill-effects in those who are *not* insufficient. But screening *all* women for vitamin D is expensive, and we need agreed definitions and thresholds for this to be useful.

Adequate maternal calcium intake is another important determinant of bone mineral density in childhood. For example, in India studies have shown that children whose mothers consume calcium-rich foods during pregnancy (milk, milk products, pulses, meats, green leafy vegetables and fruit) have higher bone mineral content and bone-mass density at six years of age.[18] Calcium supplementation in pregnancy can improve foetal bone mineralisation in women with otherwise low dietary calcium intake, but that this has little effect in pregnant women with adequate dietary calcium.[19]

A number of factors also influence calcium transfer across the placenta, including maternal vitamin D stores and the levels of placental calcium transporter protein 'PMCA3'. The level of this protein also favourably predicts bone density in the newborn.[20] Other proteins expressed in the placenta also appear to predict foetal bone mineralisation. This has caused experts to speculate that environmental factors may alter the patterns of bone development by epigenetic effects (Chapter 11) that modify the expression of these proteins.

General nutrition, again, is also important. Under-nutrition leads to poor foetal growth – a recognised risk for reduced bone mass and reduced mechanical strength in the offspring in later life. Several metabolic hormones implicated in body composition and regulation of fat stores, including IGF-1 and leptin (Chapter 4), are important for bone mineral deposition and may mediate the effects of maternal nutritional status. In healthy pregnancies, maternal fat stores may mediate their effect on foetal bone accrual through variation in foetal leptin concentrations, which have been shown to predict both the size of the neonatal skeleton and its mineral density.[21]

Stressful maternal conditions may also change the programming of cortisol (HPA axis) stress responses in her offspring (Chapter 7), which can have further indirect effects on later bone metabolism. Cortisol is a steroid hormone, which like steroid drugs can thin the bones when levels are increased. Birth weight and weight in infancy can predict both growth hormone and cortisol levels in later adult life.[22-24] These hormones influence bone metabolism. As described in Chapter 7, stress in early life can have lasting effects on the programming of the cortisol stress responses in adult life, through epigenetic changes in gene expression in the brain.[25] The associated metabolic effects are also linked to bone health.

In animals, dietary restriction, particularly protein, reduces skeletal growth and delays maturation of bone growth plates.[26] Growth-restricted animals also show many other permanent changes in gene expression that predispose to a range of metabolic disorders and later disease effects, demonstrating likely links between bone metabolism and other metabolic processes (Chapters 4 and 5).

These observations reinforce the need for healthy dietary patterns in pregnancy. Indeed, studies have shown that mothers who have healthy general dietary patterns, with higher intakes of fruit, vegetables, and wholemeal bread, rice, and pasta and low intakes of processed foods (such as tinned food, soft drinks, potato chips and processed meats) are more likely to have children with larger bone size and bone mineral density at nine years of age.[27] This effect was over and above any influence of other factors such as childhood height, weight or exercise, and other maternal factors such as smoking or vitamin D levels. Several nutrients, including magnesium, potassium, vitamin C, vitamin K, B vitamins, and carotenoids are also more important for bone health than previously realised.[28]

There is also concern that chemicals present in packaged food and soft drinks, such as the tin-related compound stannous chloride ($SnCl2$), could have a range of adverse effects on the developing foetus. In pregnant animals, $SnCl2$ has specific effects on the developing skeleton and the immune system.[29] During early development, we are more vulnerable to the effects of chemical toxicants found commonly in our modern environment.

Many other maternal factors are important. A woman's age at conception, her general health, physical activity level, education, socioeconomic status, lifestyle (e.g. smoking), and body composition are all important in determining foetal growth and the risk of low birth weight (Chapter 4). A number of these factors are associated with bone-mass development and later fracture risk.[30]

The father counts too. Our father's skeletal size also predicts our skeletal size independently of the mother's body build.[31]

Clearly, a range of early events influence bone growth and mineralisation and determine the peak bone mass in adulthood. Maternal factors and maternal environmental exposures, in particular, have a major effect on BMD and peak bone mass as the major determinant of future osteoporosis, and provide the most obvious and cost-effective avenue for prevention.

PRETERM INFANTS ARE AT HIGH RISK OF BONE DISEASE:
Babies born before the third trimester miss out on the critical period for calcium transfer in utero. And more than 10 per cent of all babies are now born prematurely.[32]

Deprived of their intrauterine supply of calcium and phosphorus, these infants are prone to a very early form of bone demineralisation ('osteopenia'). The severity will depend on the degree of prematurity and birth weight and it is common in infants born before twenty-eight weeks, affecting around 25 per cent of infants who weigh less than 1,500 g and over 55 per cent who weigh less than 1 kg. This form of osteopenia may manifest in multiple fractures or overt rickets, with deformities of the skull and thickening of the ribs and the wrists. Some do not manifest, but are diagnosed by low serum levels of phosphorus and abnormal blood markers of bone mineralisation. X-ray examinations also often reveal bone abnormalities.

It is now routine practice to supplement these infants with calcium and vitamin D, and the short-term benefit on bone mineral density is obvious; but, in the long term, these effects seem to disappear. Older studies, before routine supplementation, show that bone mineral content is 36 per cent higher at three months of age in supplemented than non-supplemented preterm infants, but by ten years of age there were no longer any differences.[33] The optimal length, quantity and method of mineral supplements in preterm babies are still unclear, and the need for high calcium and phosphorus intakes after hospital discharge is still controversial.

Some studies show that breastfeeding has a greater effect than supplementation, despite the fact that human milk has a low mineral content.[34] This may be due to still poorly understood non-nutritive metabolic or anti-inflammatory effects of breastmilk, and still needs to be elucidated. In full-term babies, breastfeeding does not appear to have any major effects on bone development.[35]

Although osteopenia of prematurity generally has a good prognosis, and preterm infants show 'catch-up' mineralisation in their first year of life, children and young adults born preterm are generally shorter, lighter, and have lower bone mass than their full-term peers many years later (although their bone mass is still proportional to their height and lean mass).[36, 37] There is speculation that the use of steroid medications (such as dexamethasone, a synthetic steroid that is twenty-five times more potent than the cortisol produced in our bodies) may have effects on bone formation in addition to other metabolic effects on body composition.

Extremely low birth-weight preterm infants rarely survived in previous generations, so any long-term studies are recent. These studies suggest that premature infants may be more prone to a number of adult onset NCDs, including heart disease and hypertension,[38] and have differences in the heart muscle that may predispose to heart disease.[39] They also have a greater percentage of body fat and less lean body mass[40] with increased risk of obesity and osteoporosis. Given the higher rates of preterm birth and improved survival, there is a pressing need to understand the long-term consequences and how these might be best prevented.

FACTORS THAT AFFECT BONE DEVELOPMENT IN CHILDHOOD:
Long-term studies have shown that reduced childhood growth is associated with increased risk of hip fracture in later life.[41] While

this may be due to a combination of factors, the findings support measures to optimise childhood growth as a preventive strategy.

Calcium requirements increase during periods of rapid bone growth in childhood and adolescence. Lower consumption of milk and dairy products in childhood and adolescence has been associated with lower BMD in adult women. When dairy products are displaced by soft drinks (cola in particular) there is also an associated fracture risk in school-age children.[42.] This has sparked some debate about whether this is due to the phosphoric acid and caffeine in these drinks increasing bone fragility, or the concomitant decline in dairy intake.

Whichever is the case, there is little doubt that the consumption of soft drinks, particularly in children and adolescents, is of substantial public-health concern. A systematic review of eighty-eight studies confirmed that soft drink intake was associated with lower intakes of milk, calcium, and other nutrients.[43] Some studies also showed reduced serum calcium levels according to the level of soft-drink consumption. Notably, studies funded by the food industry were less likely to find adverse effects than non-industry-funded studies.[44]

The increased caloric consumption from soft drinks and other unhealthy foods also increases body fat. Recent body composition studies in 8- to 10-year-old children confirm that for each kilogram of 'lean' body mass there is a 28 g increase in total bone mineral content, compared to only a 9 g increase per kilogram of fat mass.[45] Similar effects are seen with total bone area and BMD, indicating that increasing *lean* mass in children may help optimise bone acquisition and prevent future osteoporosis.

Physical inactivity is also a major risk factor for osteoporosis. Exercise and weight-bearing activities are important in skeletal development, bone-mass accrual, bone structure and strength. In later life, physical activity can also help reduce of bone loss. Exercise in children is particularly important for optimal bone development.[46] Without intervention, the currently declining physical activity levels in childhood and adolescence are likely to contribute considerably to increases the adult fracture risk in the coming years.

Physical activity influences bone health in a number of important ways. Load-bearing activities appear important for building bone mass and improve health across the lifespan;[47, 48] however, the gains in bone strength are greatest in pre-pubertal and pubertal children than for any other age. Only from childhood do the benefits persist, with gains in bone mass and structural adaptations maintained through puberty and into adulthood, even after exercise programs have ceased. Several studies have shown that even short-term bouts of load-bearing activity during childhood can have a lasting effect, well after cessation of the loading activity.[49, 50]

While high-impact activity may be good for bone strength, there is some evidence linking youth sports injuries, particularly acute injury of the knee and ankle, to osteoarthritis (OA) later in life (below).[51] This highlights the need for a balanced approach that reduces osteoporosis without adding to the risk of OA.

Our bones adapt to the forces we experience by increasing mass and remodelling to increase strength accordingly. Children who spend more time in active play have improved bone structure compared to less active peers. Studies indicate that children (aged five to eleven years) who participate in forty minutes of at least moderate-intensity physical activity per day are expected to have 3 to 5 per cent greater measures of axial and bending strength than those participating in only 10 minutes per day.[52] These studies suggest that bone adapts during regular *daily* activities in childhood and that meaningful increases in bone strength are possible without extraordinary amounts of activity. Effects do also vary with other background factors.

The level of physical activity before and around the adolescent 'growth spurt' appears to be particularly important for bone-mass accrual. Nearly 40 per cent of adult bone mass is achieved during the two years before and during the two years of adolescent 'peak height velocity' (PHV) – around eleven to twelve years of age for girls and thirteen to fourteen for boys. Athletes who began sports before or around menarche have greater bone mineral content than athletes who start training after menarche.

Pre-puberty and puberty, therefore, may be another important 'window of opportunity' for bone-health programming, via exercise programs.

Overall, the research highlights the importance of establishing physical-activity habits early in life to promote lifelong skeletal health, balancing the risk of high-impact activity against the risk of later degenerative joint disease. There is little doubt that sedentary behaviour has wide-ranging adverse effects on health,[53] and that regular physical activity should be encouraged from an early age.

The impact of illness in childhood:

Premature osteoporosis and stunted growth are common complications of chronic inflammatory disease in childhood. And over the last thirty years the prevalence of chronic conditions in children and adolescents has increased significantly.

Many diseases and disorders have been associated with osteoporosis. In some cases osteoporosis is secondary to immobility, or to the increased metabolic demands of chronic inflammation. In other cases osteoporosis is due to hormonal abnormalities, nutritional deficiencies or genetic defects. Examples of hormonal abnormalities include diabetes, thyroid disease or adrenal disease. Oestrogen deficiencies in girls and testosterone deficiencies in boys will also affect bone density. Kidney diseases can lead to bone diseases as a result of associated abnormalities in metabolising calcium, phosphate and vitamin D. Conditions that cause nutritional deficiencies include inflammatory bowel disease, coeliac disease or cystic fibrosis. Anorexia and bulimia can also lead to osteoporosis. Medications used to treat chronic diseases, particularly steroids, predispose to osteoporosis. Chemotherapy and radiotherapy also interfere with bone growth and mineralisation.

Essentially, it's a very long list of conditions! Bone-health problems should be among the complications anticipated and addressed in children with chronic disease.

OUR TEETH PROVIDE IMPORTANT CLUES TO THE 'EARLY ORIGINS' STORY:

Teeth are another fascinating part of this story. They carry the history of our personal health, and by studying teeth that are *millions* of years old anthropologists have learned much about the health and development of our ancient ancestors. This has been a most unexpected source of evidence that supports Barker's 'Developmental Origins of Health and Disease' hypothesis.

Oral health shares many of the NCD risk factors such as poor diet and tobacco use.[54] Most oral diseases are caused by sugars in the diet, poor levels of oral hygiene, smoking and injuries, and there is currently an initiative to incorporate oral diseases into programs for the integrated prevention and treatment of NCDs.

Early nutrition is important for our developing teeth. And the consequences of poor dentition are often underestimated. For example, dental caries and chronic inflammation is a risk factor for coronary heart disease (Chapter 6).[55] Healthy dentition depends not only on the teeth themselves, but the health of the bone, gums and the soft tissue that support them.

The initial calcification of our primary teeth begins in the second trimester of pregnancy. For our secondary teeth it begins soon after birth. Vitamin deficiencies during this time can affect tooth structure, mineralisation, enamel formation and susceptibility to decay. As for our bones, adequate levels of calcium, phosphorus and vitamin D are essential elements in tooth formation. Fluoride makes our teeth more resistant to demineralisation and decay. Vitamin A and vitamin C are also essential for the health of the connective tissues that support our teeth. A healthy diet and life-style can generally supply all of these nutrients. The addition of fluoride to public water supplies has been one of the most effective methods in reducing tooth decay (though too much ingested, as opposed to applied topically, can lead to dental fluorosis, or staining of the teeth).

And then there is sugar. Fruit juices, soft drinks, and other sweet things promote tooth decay and enamel loss. The naturally occurring bacteria in our mouths ferment sugars to produce acid (lactic acid), which gradually erodes the enamel and allows the bacteria to penetrate more deeply into the tooth.

But teeth are also important to the 'early origins of health' story in other ways. Anthropologist Professor George Armelagos has been studying teeth from our ancestors for many years.[56] Teeth from the Australopithecines living in the South African Pleistocene (1 to 2 million years ago) have revealed that enamel defects in early life predicted earlier death.[57] The same relationships have been seen in other ancient populations. Enamel is secreted in a regular, ring-like fashion from the second trimester of foetal development, so disruptions caused by disease, poor diet or psychological stress, will show up as grooves on the tooth surface.[58] This 'snapshot' of life health can be compared to the skeletal evidence, which gives the age at death.[59] This reveals that people who developed tooth-enamel defects in foetal life or early childhood were more likely to die twelve to fifteen years earlier than those with no enamel defects.[60]

The teeth provide a *measure* that stressful events early in development are linked to shorter life spans. Bad teeth did not *directly cause* the shorter lifespan, although they would have contributed. Early-life stressors would have taken their toll on other developing systems as well as teeth, to reduce life expectancy.

THE BENEFITS OF EARLY SCREENING AND INTERVENTION FOR
SKELETAL DEFECTS:

For hundreds of years the cornerstone of orthopaedics (treating musculoskeletal conditions) has been the correction and prevention of spinal, bone and joint deformities in children – through both surgical and nonsurgical treatments. 'Orthopaedics' literally means *straightening children*. Early detection and early intervention for skeletal abnormalities can make a dramatic and permanent difference.

A very good example of this is early screening for congenital hip dislocation – 'developmental dysplasia' of the hip (DDH). In this condition, abnormal development of the hip socket (acetabulum) and the upper thigh bone (femur) lead to instability of the hip joint. Newborns often have greater laxity of ligaments around the hip. If the head to the femur slips out of its socket, and remains dislocated, the acetabulum does not develop normally, causing permanent deformity. In some cases the hip functions well for years, but gradually progresses to functional disability, pain or arthritis. In more severe cases DDH manifests much earlier, with delay in walking, differences in leg length, arthritis in the abnormal joint, and other joint strain. Back pain and spinal deformities (scoliosis) are also common complications. All this can usually be prevented with simple, early treatment. In most countries paediatricians examine for DDH as part of routine 'normal baby checks' making sure the hip is moving normally, and that the femur does not dislocate easily. An ultrasound exam can confirm any diagnosis. Affected infants are then usually placed in a flexible ('Pavlik') harness to hold their hip in place until it is developing normally. This is highly effective during the first six months of life. Older children may require more invasive surgical management, but the prognosis is generally very good.

Paediatricians also screen for other skeletal deformities, including foot deformities and lateral (sideways) curvature of the spine (scoliosis). 'Club foot' usually resolves with age without intervention. Scoliosis in the newborn usually indicates underlying defects in the vertebral bones and other malformations. In older children (more often girls) the underlying cause of spinal curvature is not clear. Mild spinal asymmetry is common and often of no significance, but severe curvature may progress with age – although this is difficult to accurately predict. Progression appears more likely during the adolescent growth spurt. Although most people with scoliosis do not have increased pain or disability, significant deformity can be an issue. Scoliosis can be detected visually, and x-rays will reveal the severity of scoliosis. Bracing does not *reverse* an existing curvature, but it may prevent curve progression before maturity and growth is complete; this may reduce the need for surgery, though there's no consensus on this.

Schools in many regions have scoliosis screening programs, around the age of puberty, but their effectiveness is not proven. Spinal examination is also often part of the routine health checks of children. In some regions, such as Australia, routine screening has been abandoned in favour of National Self-Detection Programs which encourage girls and their parents to assess for this condition and to have family doctors assume care of minor curvatures.

Other apparent 'deformities' such as bow-legs, knock-knees or in-toeing (pigeon toes) are usually positional variants that will resolve with normal growth. The confines of the womb predispose baby to several positional rotational deformities, for example 'club foot' when the forefoot is turned inwards, which resolves with age in 90 per cent of cases. In-toeing and out-toeing also resolve without intervention, and surgical treatment is rare. Older infants and toddlers often exhibit 'bowing' of the legs as they start to walk, then go through a period of 'knock knees' between the ages of two and five. These developmental transitions progress to normal adult alignment and intervention is only needed if there is abnormal persistence or underlying pathologies.

A number of joint diseases in children can lead to inflammation, joint or cartilage damage, and early osteoarthritis (OA). A common hip disorder of adolescence known as slipped capital femoral epiphysis (SCFE) occurs when the head of the femur (the 'capital') no longer sits squarely on the neck of the femur. Obesity is a major risk factor for this disorder, which is essentially a fracture through the growth plate caused by shearing forces. Over half of those affected are in or above the 90th percentile for weight. It is most common in growth spurts (typically between ten and sixteen years) and is twice as common in males. SCFE typically manifests with an insidious onset of thigh or knee pain and a painful limp. This is an orthopaedic emergency. Slippage at this growth plate can cut off the blood supply to the head of the femur and cause the bone to die. The head must be surgically pinned in place to prevent further slippage, followed by a period of strict bed rest.

Patients with SCFE are at increased risk for premature degenerative arthritis or osteo arthritis (OA). In fact, unrecognised SCFE is believed to be a common cause of OA even in the absence of complications. Rising rates of adolescent obesity will likely contribute to an increased risk of OA in later life for the latest generation.

A number of other conditions and treatments can lead to bone death or 'avascular necrosis', including chronic diseases such as renal failure or sickle-cell anaemia, and medications such as steroids.

An idiopathic form (meaning no obvious cause) of avascular necrosis known as Perthes' disease (or Legg-Calve-Perthes' disease) is another common cause of hip disease which typically occurs between four and eight years of age. In this disease, the head of the femur dies and stops growing, but healing eventually occurs as new blood vessels infiltrate the dead bone. There is still some loss of bone mass and some degree of deformity: x-rays usually show a flattened or misshapen head of the femur bone. The degree of deformity as the disease heals is a determinant of long-term disability and degenerative OA. Children under the age of eight years are less likely to have severe deformities and are preferably treated with traction and braces until the disease has run its course and healing has taken place. Physiotherapy and low-impact exercise such as swimming are important in achieving a smoother interface between the femur and hip socket. After the age of ten years, the risk of developing OA is considerable, possibly because of the reduced plasticity to remould bone that comes with age, and older children may require more aggressive treatment. There is no 'cure' for Perthes' disease, and because it is self-limiting, treatment is focused on minimising damage during the vulnerable period.

The incidence of Perthes' disease varies between and within countries. While some of this variation appears to be due to racial differences, genetic factors do not fully explain these wide variations. Many, but not all, studies suggest a steep disease gradient across social classes, with higher rates in the less privileged. Latitude is another strong association, possibly implicating lower sunlight exposure as a risk factor.

These observations suggest that the early environment may play an important role. A twin with lower birth weight is more likely to develop Perthes' disease than the other twin,[61] suggesting that factors associated with low birthweight are also involved in the development of Perthes' disease. Children with Perthes' are also generally shorter than their peers. A greater propensity for other congenital anomalies also points to intrauterine effects.[62] In nations such as the UK, there is a very strong linear relationship with poverty, suggesting that this disease is closely linked to childhood deprivation, poor nutrition and parental smoking.[63] Children who develop Perthes' disease also have a much higher risk for developing cardiovascular disease and hypertension in later life, possibly related to early abnormalities in local vascular function. All this indicates that Perthes' shares many risk factors with other NCDs, including poor intrauterine growth.

Together, these observations reveal that many conditions in early childhood may predispose to a much higher risk of degenerative joint disease in later life. Nutrition, body weight, physical activity and other lifestyle factors play an important role in determining this risk. A range of congenital skeletal abnormalities can also contribute to both short-term and long-term health, and early intervention can improve outcomes in many cases.

POSTURE AND BEHAVIOUR IN CHILDREN – LAPTOPS, TABLETS AND COMPUTERS:

Born into a world of accelerating technology, childhood activities and behaviour have changed radically in the last twenty years. Increasing 'screen-time' has raised concerns about children's musculoskeletal health[64] as well as the more obvious impact on the risk of obesity and associated NCDs.

By 2002 over 1 billion personal computers were reported to have been produced, and there may be more than 2 billion devices in use in 2014. In Australia, 86 per cent of households with children had home internet access in 2009, and one in six children between eight and fourteen years of age had a computer in their bedroom.[65] The time spent online increases with age, and in 2009 one third of 12- to 14-year-olds spent ten hours or more online each week, compared with 5 per cent of 5- to 8-year-olds. Although there are regional variations, this reflects general global trends. It certainly seems to be my son's preferred mode of communication!

Several studies have reported that children using computers may be at risk of developing musculoskeletal symptoms.[66] Computer use may attribute to the increase rate of neck–shoulder pain and low-back pain among adolescents since in the 1990s; neck pain is particularly common with computer use, especially with daily use of more than three hours.[67] Several sources indicate around 60 per cent of children and adolescents report discomfort whilst using their laptop or other devices.[68] A similar proportion experience aches and pains while carrying a laptop.

Computer use has also been associated with changes in habitual spinal postures in adolescents.[69] This does depend on gender, with males more likely to show increased head and neck flexion, and females increased lumbar lordosis (swaying of the lower back).

Although guidelines have been developed to address the potential effects of computer use in the workplace and at home, newer forms of technology such as smart phones and tablet computers have not been well studied. Published guidelines may also not apply as well to children, who adopt different postures than adults while using computers.

In my hometown, Perth, Professor Leon Straker and colleagues have been studying postural changes in children with computer use for some years. They found that children using tablet computers are more likely to have elevated shoulders, asymmetrical forward hunching of the trunk, and more tense muscle activity around the neck, than if they are using a desktop computer.[70] However, greater variation of both posture and muscle activity could offset the negative effects of more static postures with the use of a desktop computer. That said, the effects of tablet use were actually similar to paper use. And traditional classroom activities are also associated with static postures.

There are also potential adverse effects of carrying heavy school bags. Despite having a laptop for school, my son's school bag is just as heavy as the one I had to carry in the 1970s full of weighty books and papers. Today, padded backpacks with waist straps help distribute the weight. Even so, a child should not carry a bag that weighs more than 10 per cent of their bodyweight. And bags need to be packed and worn properly to minimise injury.

THE DANGERS OF SITTING:

There has been much focus on 'lack of exercise', but 'time spent sitting' may be an independent risk factor for disease.

Prolonged sitting has long been considered detrimental to health, but newer studies show that this is a risk factor for 'all-cause mortality', independent of physical activity.[70, 71] In an Australian study, 'sitting time' was a consistent risk across the sexes, age groups, body-mass-index categories, and physical activity levels. It was seen for *both* healthy people and people with pre-existing cardiovascular disease or diabetes. In Norway, similar observations have been made.[72] These studies focused on adults, but it is also important in children for health, body weight and longer-term behavioural habits. Public-health programs should specifically target sitting time as well as physical activity levels.

A recent systematic review combined data from 983,840 school-aged children and adolescents between five and seventeen years and found that watching TV for more than two hours each day was associated with unfavourable body composition, decreased fitness, and lowered self-esteem. Sedentary behaviour was also associated with reduced social behaviour and decreased academic achievement. Interventions to reduce sedentary time significantly reduced body mass index (BMI).[73]

Decreasing any type of sedentary time will improve health and reduce health risk in school-aged children. While substantive effects may be the result of reducing BMI, there are other metabolic, musculoskeletal and physiological benefits from reducing sedentary activity.

INFLAMMATORY JOINT DISEASE CAN ALSO BEGIN IN EARLY CHILDHOOD:

'Rheumatoid' conditions include immune conditions that affect our joints, connective tissues, muscle, tendons, and other body tissues. These inflammatory joint diseases are among the many 'autoimmune' conditions, like type 1 diabetes (Chapter 5), which are caused by an immune attack against our own tissues. Arthritis and joint destruction occurs when this immune assault is directed at joints and connective tissues, as seen in rheumatoid arthritis (RA) and systemic 'lupus' erythematosus (SLE). Inappropriate immune responses can be initiated at any age, and the first signs of inflammatory joint disease can begin in early childhood. Chronic inflammatory joint disease can also increase the risk of secondary OA and osteoporosis through the inflammatory disease itself, the medications used to treat it, or reduced ability to weight bear or exercise.

In children, autoimmune arthritis is generally referred to as 'juvenile idiopathic arthritis' (JIA). Early-onset JIA occurs in preschool children, and is much more common in girls. Boys tend to be around ten at onset of the disease.

The prevalence of JIA is somewhere around one to four per 1,000 children, although there are wide regional variations.[74] Although this condition can remit and even resolve in some cases, there is also potential for longer-term inflammation and complications that may have a lasting impact on function, growth and quality of life. Early diagnosis and management are important for reducing the impact of the disease and optimising both immediate and long-term quality of life.

JIA can take several forms. It can affect multiple joints in a symmetrical pattern ('polyarticular' disease) or it can affect only a few (often larger joints) in an asymmetrical pattern ('oligoarticular' disease). Some children also have more generalised disease with fevers, rashes and inflammation of internal organs such as the liver and spleen ('systemic' disease).

Although JIA is defined as 'idiopathic' because the cause is unknown, it is clear that both genetic and early environmental factors contribute. The same is true of adult-onset rheumatoid diseases such as RA or SLE where autoantibodies, for example rheumatoid factor (RF), are detectable in blood years before symptoms begin. This suggests they are initiated much earlier in life, as in other autoimmune conditions.

Although there is a well-recognised family tendency for immune disease such as RA, SLE and JIA, environmental factors are important in initiating, precipitating and exacerbating these conditions.[75] Genetic variants (e.g. of HLA-DRB1) carry a greater risk of developing some forms of RA, but not everyone with these genes will develop the disease. Only 15 per cent of twins are 'concordant' for RA; in the vast majority of cases only *one* twin has the disease.

Factors that have effects on the developing immune system (and increased risk of RA and JIA) include patterns of microbial exposures, breastfeeding and nutritional factors as well as smoking and other environmental toxins.

An analysis of more than sixteen studies has shown an increased rate of RA in current or past smokers.[76] There is also evidence that smoking induces a specific class of autoantibody in genetically prone individuals.[77] It may chemically modify tissue proteins (citrullination) and increase the risk of antibodies to the modified protein.[78] Maternal smoking in pregnancy increases the risk of both RA and JIA, particularly in girls. In Finland, researchers found that girls whose mothers smoked ten or more cigarettes per day during pregnancy were three times more likely to develop JIA.[79]

Infectious agents are also implicated. Earlier studies suggested seasonal variations with the onset of systemic JIA, implicating viral infections as a triggering event. However, this finding is not consistent in other populations. Children hospitalised for any infection during their first year of life have been shown to have an increased risk of later RA, although this was only seen for certain subtypes of the disease.[80] On the other hand, more hygienic conditions in childhood have also been associated with an increased risk of RA in later life.[81]

Together these findings *might* suggest that generally cleaner environments in early life could alter immune maturation in ways that increase the risk of autoimmune diseases, but that specific infections might still play a role in triggering the disease. Of course, none of these associations prove any causal links. But it is interesting that the same kinds of associations are seen with other immune diseases such as type 1 diabetes (Chapter 5) and allergy (Chapter 10). In other words, the level of biodiversity and the types of microbial exposure may be as important as the number of infectious episodes.

Another factor of growing interest is vitamin D, which has been shown to modulate the immune system and is particularly important for the regulation of the 'self-tolerance' mechanisms that inhibit autoimmunity. There are studies showing that the collective risk of one or more autoimmune disease (including RA and SLE) varies with the season of birth and early patterns of sunlight exposure; however, there are lots of factors, including patterns of viral infections, that could explain a seasonal pattern.

While the relationship with vitamin D and *some* autoimmune diseases is reasonably convincing, for specific *joint* diseases such as JIA, SLE and RA, the relationship is less clear. Although one study of 29,368 women noted that RA was more likely in those who had lower dietary vitamin D intakes,[82] this has been criticised because sunlight exposure (a more important determinant of vitamin D) and serum levels were not measured.[83] Others have found no relationship between vitamin D status and the risk of SLE or RA.[84]

Randomised control trials are the best way to examine these relationships, but these are difficult and expensive. The next best way is by studying populations where interventions have been undertaken as part of public-health policy. In Finland, a researcher took this opportunity to compare children who regularly took the recommended 2,000 international units (IU) of vitamin D each day with those that did not.[85] In over 10,000 children they found that the children who took regular vitamin D were substantially less likely to develop autoimmune (type 1) diabetes. So far, the same relationship between early-life supplementation and RA or JIA risk, though plausible, has not been demonstrated.

Obesity and 'over-nutrition' have been extensively examined in relation to RA and other immune diseases. Adipocytes increase the secretion of inflammatory cytokines in the circulation. Obesity increases the amount of oestrogen in the circulation, which may affect immune function, and is consistent with the higher risk of RA in females. Other hormones involved in appetite regulation (such as leptin, Chapter 4) have effects on immune function, and may also contribute to the greater risk of RA in obesity in some but not all studies. Larger babies have been shown to be at increased risk of both allergy and autoimmunity (RA). The reasons for this are not clear, but could be related to other lifestyle factors associated with increasing economic prosperity.

The dietary elements that have been examined in relation to RA range from omega-3 polyunsaturated fatty acids (PUFA), antioxidants, red meat, alcohol, and caffeine.[86] In short, the evidence for most of these is conflicting and inconsistent. Perhaps the only nutrients that have been shown to be beneficial are the omega-3 PUFA, largely in the treatment of existing rheumatoid disease. The role in prevention is less clear. In Western societies, a higher level of fish consumption has also been shown to have a protective relationship against developing RA.[87–89] This is consistent with the immune effects of omega-3 PUFAs[90] and protective relationships seen using maternal supplementation with fish oil in pregnancy and the reduction in other inflammatory conditions such as food allergy and eczema.[91,92]

Finally, breastfeeding has recognised immunomodulatory properties. But while some studies have shown associations between breastfeeding and reduced risk of later RA,[93] others have not.[94]

In summary, many different factors are likely to initiate and drive abnormal autoimmune responses. The very fact that these immune responses can *predate* clinical disease by years clearly implicates the early environment in many cases, particularly as genetic factors cannot fully explain these conditions. Several early-life factors relate to the risk of RA, such as cigarette smoke and high birth weight, and for others there is a high index of suspicion, such as vitamin D insufficiency and patterns of microbial exposure.[95] Shared risk factors with other autoimmune and allergic immune diseases (and other NCDs) reinforce the likely impact of modern environmental changes on the immune system. On the other hand, RA has followed different patterns with time, geography and migration compared with some other autoimmune and allergic conditions, reflecting the obvious differences in the gene–environmental interactions that lead to specific conditions.

THE EARLY ORIGINS OF OSTEOARTHRITIS:

Although OA was previously thought to be a normal consequence of ageing, it is now understood to be the result of more complex interplay between joint mechanics, local and systemic inflammation, genetics and a range of biochemical processes. OA has been generally classified as 'degenerative' but many studies have shown some degree of inflammation in both early and later stages of the disease. And, again, genes only account for a small proportion of the overall risk.

The events that lead to this disease vary widely between individuals, and are still incompletely understood. A number of associated 'connective' tissues are affected, including the articular cartilage, subchondral bone, menisci, and the soft tissues around the joint including the synovial membranes. In OA, degradation exceeds the regenerative repair processes, leading to progressive joint destruction. Inflammation and changes in cartilage metabolism, with disturbance in the remodelling processes, are integral to the disease.

Age is still the strongest association with OA. It is estimated that around 80 per cent of people over fifty-five have some degree of OA, reflecting mechanical 'wear and tear' on joints, and cartilage degeneration. Other mechanical risk factors include previous injuries, joint mal-alignment, or malformations. Childhood disorders, such as developmental dysplasia of the hip (DDH), SCFE or Perthes' diseases are also significant risk factors, as discussed above. Females are at greater risk of developing OA and experiencing more rapid joint destruction, which may relate to hormonal differences. Some studies suggest that postmenopausal hormone replacement therapy (HRT) protects women from developing large-joint OA, but other studies show no effect.

Obesity may lead to increased mechanical stress on joints, particularly in the knees, but is also associated with a significantly higher risk of OA in non-weight-bearing joints such as the hands, suggesting that other mechanism may also be at play. Animal models indicate that the biochemical changes associated with obesity accelerate OA beyond the effect of increased joint 'load' of increased body weight.[96] In these animals, adipose-associated inflammation appeared to have a role in the development of obesity-induced OA.

Some of this risk may lie in the fact that obesity is a chronic inflammatory state (Chapter 4).[97] It has been proposed that associated low-grade systemic inflammation may also increase the risk of both OA and osteoporosis.[98] Weight reduction is likely to protect against well-recognised musculoskeletal complications. Importantly, all of these benefits will be greater if obesity is addressed at an early age, or prevented in the first place.

Chronic low-grade systemic inflammation even in the absence of obesity or other overt disease may also increase the risk of osteoporosis, and possibly OA. An increasing tendency for chronic low-grade inflammation, as detected by higher inflammatory markers such as C-reactive protein (CRP), has been associated with osteoporosis (nontraumatic fractures), suggesting effects of chronic low-grade inflammation on bone quality.[99] This propensity for inflammation is programmed to some extent in early life (Chapter 6).[100]

Exercise and sports can be either preventive or a risk factor for OA.[101-103] The bottom line is:

- if joints are *normal,* regular exercise is *protective*, and reduces the risk of OA
- repetitive, *high-impact* exercise *does increase the risk* of OA in normal joints
- in *abnormal* joints, even repetitive, *low-impact*, recreational exercise can increase the risk of OA.

Previous injuries or joint abnormalities increase the risk of OA. Small insults over time may also be a factor, more so for OA of the knee and hand and less so for the hip. Certain activities, such as prolonged, regular squatting or kneeling (more than thirty minutes per day), or climbing more than ten flights of stairs per day may increase the risk of OA. These risks are amplified in the presence of obesity.

OVERCOMING INACTIVITY AND OUR SEDENTARY LIFESTYLE:
Physical activity and good nutrition are arguably the most important factors for musculoskeletal health and reducing the long-term burden of both OA and osteoporosis. We face complex challenges in making these apparently 'simple' changes. These lie in the structures and functions of our societies and the way that we live.

There are important social and cultural factors at play. In low- and middle-income countries, cultural influences play a significant role in disparities between the sexes, particularly during the school years.[104] In developing regions, adolescent girls are less active and place less value on participating in physical activity.[105] In some societies, cultural and economic influences curtail the physical mobility for many girls and women, such as sport being considered 'unfeminine' or restricted through dress or income. In their 'call to action', the NCD Alliance has stressed the need for health systems to be more gender-responsive and pay adequate attention to different gender needs and priorities.

In high-income countries, current health-promotion programs typically target individuals and their behaviour, but it is increasingly evident that there needs to be a wider collective social effort to change how we live, how we work, and how we move within our built environment. The layout and structure of our 'built environment' has an important role in the dynamics of our behaviour, but it is often overlooked. There are now experts that study how to build towns and cities in ways that are more conducive, or actively facilitate, healthy behaviour in the population. But economic constraints are often major obstacles to change.

We already know that *early* interventions *can* change behaviour and dietary practices. Recently, Professor Louise Baur and her team conducted a home-based intervention in socially and economically disadvantaged areas of Sydney[106] and saw that by two years of age, the intervention group had a significantly lower BMI than the control group and showed many other positive behaviours, including less TV-viewing time, and more physical activity. The size of the benefit in this study was quite large relative to the effects of other interventions in older children.[107] School-based initiatives are another important opportunity to influence behaviour through health education, activity plans and nutritional programs.[108]

• • •

Our musculoskeletal system is intimately connected with our neural networks, our metabolic systems and our other physiological processes. Early events shape the health of our bones, joints and connective tissues, and this also has implications for the health of so many other body systems. Conditions such as osteoporosis and osteoarthritis are among the most common conditions of ageing and can substantially reduce our quality of life, and even lead to our premature death. Efforts to improve musculoskeletal health from early life *must* be an integral part of strategies to reduce the burden of NCDs.

9

The developmental origins of cancer

The big 'C' word is one that no one wants to hear. It brings images of pain, suffering and loss. We all fear it, and hope that it will not happen to us. But cancer continues to be one of the leading causes of death in both developed and developing countries. Ageing populations and adverse environmental exposures and behaviours have led to increasing rates of cancer. And adult cases may be the result of accumulated exposures from earlier life.

In 2008, there were an estimated 12.7 million cancer cases and 7.6 million cancer deaths. Over 60 per cent of these deaths occurred in the low- and middle-income countries.[1] And these numbers are increasing. Breast cancer is the most frequently diagnosed cancer in women, accounting for 25 per cent of cases. For men, it is lung cancer, largely because of increased cigarette smoking in developing countries; in developed nations, prostate cancer is now the most common cause. A substantial component of cancer risk is environmental. In children, leukaemia, brain tumours and lymphomas are the most common cancers. Childhood cancers have also been increasing[2, 3] again suggesting a role of modern environmental changes. Survival rates in both adults and children are substantially lower in developing countries because of limited access to healthcare and treatment.

SOME BASICS OF CANCER AT ALL AGES:

There are hundreds of different kinds of cancers, which can affect almost any tissue. They vary considerably, but they all feature uncontrolled cell growth. The cells become 'immortalised' through continual cell division and proliferation. As they grow, these cancerous communities invade and displace normal tissues to disrupt normal structures and functions. These aggressive and destructive characteristics distinguish 'malignant' conditions from 'benign' tumours. Cancer cells produce a range of growth factors to support their own expansion, in particular vascular growth factors. They cause proliferation of blood vessels to provide themselves with nutrients, usually at the expense of other tissues; worse still, this provides passage for them to spread to distant tissues. Spread or 'metastasis', either through the blood stream or the lymph vessels, is another typical feature of many cancers. As they grow and consume nutrients, the rest of the body suffers. Without treatment, this continues and eventually leads to death.

Once there is sufficient damage to disrupt normal functions, cancers can produce *almost any* symptoms according to their location and their pattern of destruction. Depending where the tumour is, there may be difficulty breathing or swallowing, bowel obstruction, or palpable masses or lumps. Signs of cancer may only appear after metastasis – to the bones, the brain or the liver – with fractures, altered behaviour, or signs of liver failure. Some patients present with tiredness, weight loss, anaemia or fevers caused by more generalised 'systemic' inflammation. This is also common in children with leukaemia, who may also show easy bruising and painful legs among their first symptoms.

Cancer screening programs, such as those for cervical cancer and breast cancer, mean that some malignancies can be detected before any symptoms appear. Any diagnosis is confirmed by tissue sampling and microscopic evaluation, and cancers may be variously graded according to their microscopic appearance, their degree of local invasion, or metastatic spread. In more difficult cases, analysis of metastatic lesions can reveal cells virtually unrecognisable as any mature tissue, with no obvious primary tumour and no clue of which tissue they came from originally. Fortunately, this is fairly uncommon.

Once a diagnosis is made, treatment can begin. But in many cases this just delays the inevitable. New and more-effective treatments are vitally important, but it would, of course, be far better to prevent cancer in the first place. And that is what I want to focus on here.

The stepwise processes of lethal transformation:
Malignant transformation is a highly complex and highly variable process that occurs in many steps. Usually, cell proliferation is preceded by the activation of a series of genes that are normally present but under tight control. There are two main ways this happens. First, through *increased* expression of genes that promote cell growth and reproduction. These are called 'oncogenes'. Or second, through *decreased* expression of genes that inhibit cell growth, called 'tumour-suppressor genes'. Usually both processes occur. Cancer cells usually show abnormal expression of *multiple* genes, which amplify unregulated cell proliferation.

Either mutations or increased expression can transform *normal, essential* proto-oncogenes into more hazardous oncogenes. There are also 'viral oncogenes' – pieces of viral DNA inserted into the human gene sequence – including human papilloma virus (HPV) in carcinoma of the cervix, hepatitis viruses B and C in liver cancer, and the Epstein Barr virus (EBV) in a range of lymphomas and other malignancies. Some of these effects are *direct* by inserting and expressing viral genes onto cells, but many effects are *indirect* by causing chronic infection and inflammation in the tissue (such as with chronic hepatitis).

Oncogenes all promote cell survival and more rapid turnover of cells. But to produce cancer, oncogenes usually need additional steps, such as changes in tumour-suppressor genes or in other proto-oncogenes.

One of the important tumour-suppressor genes is called protein 53 or just 'p53'. It has been described variously as the *'guardian of the genome'* and the *'policeman of the oncogenes'*[4, 5] and even earned the title *'molecule of the year'* by *Science* magazine in 1993.[6] In healthy cells, the p53 protein maintains the steps leading to cell growth, halts abnormal growth and prevents cancer. However, even one small change in one of p53's 393 amino acids can allow a cancer to grow. Although p53 abnormalities may occur in up to 50 per cent of cancers,[7] there are variations in the frequency and type of mutations depending on the form and stage of the cancer.[8] The changes in p53 may predate the cancer by many years or may occur later in cancer development, with aggressive tumour behaviour and poor prognosis. The frequent association of p53 defects with so many forms of cancer, however, illustrates the critical role of tumour-suppressor genes in preventing cancerous changes in normal cells.

Once these genetic changes (in oncogenes and tumour-suppressor genes) gather momentum, they can quickly lead to a chain reaction. Rapid cell turnover increases the chance of *additional* genetic errors, escalating further when mutations occur in the 'error-correcting' DNA repair machinery. When tumour-suppressor genes or other normal regulatory genes are lost or inhibited, the risk of malignant transformation surges. As tumours continue to evolve, they may acquire or lose other genes that allow them to become even more aggressive.[9] This can include genetic changes that allow them to metastasise and colonise new organs, evade predation by the immune system, or to send error-causing signals to damage other surrounding tissues.

This highly variable multi-step process explains why the process of cancer development varies so widely between people. The processes of ongoing evolution and genetic changes within cancers also explain why they are so difficult to treat and even more difficult to cure.

THE IMMUNE SYSTEM AND CANCER SURVEILLANCE:

Genetic 'mistakes' are a real and constant hazard, every time our cells divide. And they are actually quite common. We rely on a series of safeguards designed to protect us against this. Within the cell nucleus there are inbuilt mechanisms that constantly repair damaged DNA. If this fails, the cell has internal 'self-destruct' mechanisms that initiate 'programmed cell death' (known as 'apoptosis'). And if that fails, our immune system is able to detect and kill rogue cells. It is believed that we all produce *potentially* 'precancerous' cells on a fairly regular basis, but these are quickly picked off by the immune system, or other means, before they can proliferate. So, when the immune system becomes suppressed with chronic illness, various drugs or other factors (possibly even chronic stress),[10] cancers become more likely. As we age, cancers also become progressively more likely, probably due to an ageing immune system and the progressive accumulation of DNA errors over our lifetime.

GENETICS AND EPIGENETICS IN CANCER CELLS:

All cancer cells show changes in gene expression. Classically, this was explained by 'genetic' mutations from direct DNA damage, but the recent discovery of 'epigenetics' has provided an important additional mechanism (Chapter 11) and a new perspective on how environmental factors might promote malignancies.

If there are genetic 'mutations' in cancer cells these will develop a *different* gene sequence than other *normal* cells in the body. These are called 'somatic mutations', and are quite distinct from an inherited genetic disorder, because the mutations are not part of our inherited genes and *only* occur in the abnormal cells (and are passed only to their cancerous 'daughter cells'). A minority of cancers run in families because of inherited predisposition, but that is quite different (below).

There are some well-known examples of large-scale DNA derangements within some cancer cells, such as the 'Philadelphia' chromosomal defect seen in some patients with leukaemia.[11] DNA damage causes chromosomes to become abnormally fused at a characteristic location (in this case chromosomes 9 and 22), resulting in the production of an oncogene which greatly increases the risk of leukaemic transformation. As well as speeding up cell division, this also inhibits DNA repair and adds to the genetic instability of these malignant cells. Many other massive chromosomal abnormalities can occur as cells divide. Whole chromosomes can even be deleted. But more-common, smaller-scale genetic mutations in the DNA sequence can be enough to significantly change function – producing oncogenes or inhibiting deleting suppressor genes.

Some degree of DNA damage is 'normal' and is usually corrected by our DNA-repair machinery. But some environmental exposures damage DNA or interfere with cell biology. 'Carcinogens' that increase and propagate genetic changes include radiation, cigarette smoke, some viruses and funguses, and industrial chemicals. Repeated cellular injury, stress, heat or low oxygen supply to cells may also increase the risk of cancer.

While genes provide the DNA code, it is the *epigenetic* processes that govern gene expression. Epigenetic changes explain changes in gene expression without mutation in the DNA sequence. In fact, epigenetic changes appear to be much more common in many cancers than genetic mutations, although both are usually present.

As discussed in Chapter 11, a number of processes control which genes are switched 'on' or 'off' in any given cell. This includes the addition of 'methyl' groups that lock genes in the 'off' position and their removal to activate gene expression. Other epigenetic processes will open or close the chromosomes to control DNA transcription. When genes are 'on' they can be 'transcribed' into RNA, which then gets 'translated' into functional proteins. But active DNA can also produce small transcripts (microRNAs) that act as epigenetic regulators. These microRNAs don't produce proteins, but instead play a role in controlling protein-coding genes to reduce their expression. This provides yet another level of control to prevent aberrant gene expression and malignant transformation.

Patterns of epigenetic regulation vary widely across the genome in any particular cell, at any particular time, switching some genes on and others off. If this pattern is disrupted, so is the function of the cell. Typically, anything that reduces methylation (or targeted microRNAs) at the site of oncogenes will allow them to escape control. If there is over-methylation of tumour-suppressor genes, this will also add to the risk of malignant transformation, without the need for specific mutations in these genes.

For example, in bowel cancers a large numbers of genes show abnormal methylation,[12] including coding for oncogenes and microRNAs that normally act as tumour suppressors. When microRNAs are silenced by over-methylation it allows the *hundreds* of the genes that they normally inhibit to escape suppression. Samples from one study showed a suppressor gene (miR-137) over-methylated in 81 per cent of colon cancers compared with 14 per cent of the adjacent normal bowel tissues.[13] Silencing of this single gene can change expression of more than 400 other genes. And numerous other microRNAs may be affected in the same way at the same time, together with numerous other genes that control cell function.

Epigenetic changes also promote genetic instability and associated genetic mutations. For example, epigenetic alterations in DNA repair genes substantially increase mutation rates and cancer risk.[14, 15] Such events are likely to be of major significance in the early stages of development in many cancers. While these many epigenetic changes are potentially reversible, they tend to persist as 'epimutations' through cell divisions in cancerous clones, unless there are strong signals to change.

Epigenetic alterations provide a more dynamic mechanistic link between many environmental and dietary exposures and cancer development. External exposures modulate epigenetic changes in gene expression, largely by design to allow metabolic or physiologic adaptation to our environment. However, some contribute to cancer risk. Because epigenetic changes are inherently *reversible*, 'epimutations' provide a new therapeutic target in cancer treatment. Animal experiments manipulating epigenetic control pathways do provide support for this concept,[16] but this is a very new area of research.

All this makes it clear that cancer is the *cumulative result* of events over many years. Our lifestyles and environmental exposures contribute to a succession of epigenetic changes in vulnerable cell populations. This can lead to a 'precancerous state' in tissues, making them more susceptible to mutations or additional epigenetic changes. Our exposures much earlier in life may be important in determining later risk, creating dormant vulnerabilities or growth and development risks to genetic or epigenetic programs. But before we explore the role of the environment, let us consider inherited predisposition to cancers.

ARE SOME CANCERS INHERITED?

It has been estimated that only 3 to 10 per cent of cancers are due to a familial genetic mutation, and that less than 0.3 per cent of the population carry an inherited mutation that has a large effect on their cancer risk .[17]

One common 'cancer gene' that can 'run' in families is the inherited mutation in the *BRCA1* and *BRCA2* genes. This confers a substantial predisposition (more than 75 per cent risk) to breast cancer and ovarian cancer[18] which appears earlier than is typical. Familial genes are also implicated in 5 per cent of all bowel cancers which, again, often appear early. The most common genetic form of bowel cancer is Lynch syndrome, with an 80 per cent risk of cancer, often with onset before forty years. An even less common inherited risk (at 1 per cent of bowel cancers) is a condition caused by mutations in a 'tumour-suppressor gene' called *APC* that leads to the formation of thousands of bowel polyps (adenopolyposis) at an early age. Although these lesions are initially benign, there is a very high rate of cancerous transformation. People in families with these known 'cancer genes' need to be screened and monitored, and may choose prophylactic surgery to remove the tissue of risk.

An ever rarer inheritable condition known as Li-Fraumeni syndrome is characterised by mutations in the major tumour suppressor gene p53, and affected families are devastated with cancer. But although the p53 defect initiates the cancer formation quite early in life, malignancies may not appear for ten to thirty years.[19] And it is perhaps surprising that these people do not develop more cancers, given that the inherited defect is carried in every tissue. The p53 mutation initiates the process, but it is not sufficient by itself.

Other examples of cancer-associated syndromes caused by inherited tumour-suppressor gene mutations include retinoblastoma (tumours of the retina caused by *Rb* mutations), neurofibromatosis (caused by NF1 mutations) and familial adenopolyposis (caused by the above-mentioned *APC* gene mutations).

There are also a number of rare inherited disorders that affect the DNA repair systems, causing high susceptibility to cancers and other abnormalities of the skin, bones, nervous system, immune system and other organs. In the case of inherited disorders that affect immune function, serious infection is the more imminent threat, but the risk of cancers, particularly lymphomas, increases with age. Such significant disorders are inevitably diagnosed early and, when appropriate, treated with bone-marrow transplantation.

While genes with a 'strong effect' on cancer risk are relatively uncommon, genes with a 'weak effect' on cancer risk may be more common. An example of this is that some inherited variants might make one person more susceptible than another to the effects of cigarette smoking or an unhealthy lifestyle.

EARLY ORIGINS – WHEN DOES THE PROCESS BEGIN?

The same kinds of complex gene–environment interactions we see for other NCDs also apply to cancer risk. Many exposures begin to have an effect on our gene expression from conception, and even before – given that the egg that we were fertilised from came into being inside our mother when *she* was still in our *grandmother's* womb! In animals, environmental effects on 'grandparents' can be transmitted by epigenetic changes, though this is less clear in humans. We know that our epigenetic program is particularly vulnerable to environmental influences during early embryonic, foetal and infant development. And our environment will determine our 'lifelong editing' of early-life epigenetic memories.[20]

Early events may result in *dormant* changes that predispose to disease states in later life. Clear examples show how a mother's exposures in pregnancy influences the risk of cancer in her children many decades later,[21, 22] as we will see in a moment. These examples illustrate the possibility that some of the environmental risk in cancers may be initiated quite early, and not manifest for many years.

The concept of 'developmental origins' provides a new perspective on the role of gene–environment interactions in the evolution of cancers. It is now proposed that some *'environmental exposures during development increase susceptibility to cancer in adulthood, not by inducing genetic mutations, but by reprogramming the epigenome.*[23] This does not preclude secondary changes in DNA, but argues that the initiating events lie in environment-induced epigenetic changes *rather* than changes in the DNA sequence. And for many 'risk factors' this makes more sense, because the epigenome is *designed* to be responsive to the environment, and is much easier to change.

Between 1948 and 1971, the synthetic oestrogen diethylstilbestrol (DES) was administered to pregnant women to prevent miscarriages. This was continued until it was noticed that the *daughters* of these women began to develop gross anomalies of the uterus, vagina and other elements of the genital tract – and a previously rare form of cancer of the vagina and cervix.[24] It is estimated that in the United States alone, more than 4 million mothers and their foetuses had substantial exposure to these carcinogenic oestrogens.[25] Animal studies confirmed that DES induced very similar long-term effects in mice as in the young women.[26] And they also began to see effects in male mice that had been exposed in utero.[27] The daughters of women exposed in pregnancy also had a higher than expected rate of breast cancer after the age of forty years.[28] On a related note, there remains strong speculation that the rise in male reproductive cancers and abnormalities in male genital tract may be related to an increase in environmental oestrogens in the last fifty years.[29]

The discovery of DES effects contradicted several of the prevailing concepts at the time. The recognition that *normal* genes (without mutations) can cause cancer if expressed in the wrong place and at the wrong time was new. The concept that development was fluid and not determined simply by gene sequence was also new. These new observations revealed the plasticity of early development and that a single 'genotype' (the program) could produce diverse 'phenotypes' (the product of what we become) depending on the early environment. And all of this will also influence our vulnerability to malignancy.

ENDOCRINE DISRUPTORS IN EARLY LIFE: A ROLE IN INCREASING RATES OF BREAST AND PROSTATE CANCERS?

For many years there has been concern that environmental chemicals may be acting as 'hormonal imposters', including a large range of products used in industry, plastics, pesticides and other organic pollutants.[30]

Several environmental 'endocrine disrupter' chemicals have been shown to have similar effects to DES, including BPA, a ubiquitous environmental chemical found in plastics that has oestrogenic properties.[31] In animal studies, foetal exposure to BPA increased pre-cancerous changes and the risk of mammary cancers in adult life[32, 33] and altered the prostate epigenome during development to promote prostate disease in adulthood.[34] These effects were all seen at levels that are relevant to general human exposure, adding to speculation that the rising rates of breast, prostate and testicular cancers may be linked to the increasing exposure to 'endocrine disruptor' chemicals in the last fifty years.[35, 36]

Other endocrine disruptors with similar hormonal effects, such as phthalates, are widely used in industry (plastics, paints, and some pesticide formulations) and personal-care products (makeup, shampoo, and soaps). In-utero exposure to BPA and several phthalates (also in plastics) induces epigenetic changes that can be inherited across several generations of animals, even in the absence of ongoing exposure,[37] and can cause DNA methylation 'epimutation's in male offsprings' sperm as adults.

We see some indirect evidence of this in humans. Testosterone controls the masculinisation of external male genitalia, including penis volume, testicular size, and distance between the anus and the genitals (anogenital distance) which can all be reduced by oestrogen exposure in early life. Testicular abnormalities, reduced sperm counts, and congenital defects such as hypospadias (where the urethral opening occurs lower than usual along the underside of the shaft of the penis) are all potential results. Anogenital distance is correlated with reduced sperm counts and fertility, and has been used as a surrogate measure of hormonal exposure in foetal life when male anatomy is developing.

In recent human studies, prenatal phthalate exposure (measured in the urine of infant boys) was associated with reduced anogenital distance – which was also correlated with reduced penile volume.[38] This is consistent with the effects of phthalates in animals.[39] In some regions there has been an increase in the rates of genital defects such as hypospadias over the recent decades.[40] Hypospadias is directly associated with reduced anogenital distance, which suggests that this may be a measure of increasing exposure to putative endocrine disruptors in the environment.[41]

Men with prostate cancer tend to have a shorter anogenital distance, suggesting higher exposure to oestrogenic compounds during early development. This could be due to higher circulating oestrogens produced by the mother, or other environmental sources of oestrogens associated with a higher risk of testicular cancer.[42, 43] In adult men, levels of 'hormone imposters' (various pesticides) measured in blood also increase the risk of testicular cancers;[44] levels reflect cumulative exposure over years, and it is difficult to pinpoint the source or timing.

These relationships between 'hormone imposters' and human disorders provide a *suggestive* pattern, particularly in light of the clear effects in animals. New evidence that these endocrine disruptors can have heritable epigenetic effects adds further credibility to these longstanding concerns.

ENVIRONMENTAL RISK FACTORS FOR CANCERS – ACROSS THE LIFESPAN:

Tobacco smoking, exposure to radiation, unhealthy nutrition, obesity, inactivity, certain infections, and a range of environmental pollutants collectively contribute to about 90 to 95 per cent of our cancer risk.[45] Put another way, many cancers could be prevented by avoiding cigarette smoke, eating more fresh fruits, vegetables and whole grains, eating less meat and refined carbohydrates, regular exercise, vaccination against certain infections, moderating sunlight exposure, and maintaining a healthy bodyweight.

It is important to consider their role in early development as many of these factors have epigenetic effects and may modify patterns of gene expression in ways that alter predisposition to cancer. And because adult exposures and behaviours may be strongly influenced by our early environment. Furthermore, some specific risk factors for cancer, such as obesity, also have their origins in early life (Chapter 4).

For example, cigarette smoke is estimated to account for about 25 to 30 per cent of all deaths from cancer and 87 per cent of deaths from lung cancer.[46] For many, the damaging effects of tobacco smoke begin in childhood, or even before birth when parents smoke. Smoking rates continue to rise in developing countries.

The mechanisms of cancer risk from tobacco smoking are believed to be the result of oxidative stress, inflammation and chemical effects on many cell functions which may induce epigenetic changes.[47, 48] Smoking in pregnancy has additional effects on the developing foetus to increase the risk of chronic disease in later life. Smoking causes inflammation and oxidative stress in the placenta with associated effects on foetal growth and effects on many developing organs, including the lungs.

A recent study of over 1,000 newborns showed that maternal smoking in pregnancy induces epigenetic changes in the foetus, which can be detected as significant differences in DNA methylation in cord blood.[49] As yet, the long-term significance is not clear, but maternal smoking in pregnancy has already been associated with an increased risk of malignancies in children in a number of studies.[50-52]

Exposure to alcohol in pregnancy is a well-known cause of multiple birth defects and mental retardation, as part of foetal alcohol syndrome. In animals, gestational exposure to alcohol increases the risk of cancer in adulthood[53] and in humans there have been some reported associations between maternal alcohol consumption and the cancer (leukaemia) risk in childhood.[54] As ever, it is much harder to prove these kinds of association in humans, but we must remain suspicious when we also see effects in animals.

Many of the long-term dietary risk factors implicated in cancer risk are similar to those implicated in other NCDs. In 1981, it was estimated that approximately 30 to 35 per cent of cancer deaths in the USA were linked to diet[55] and it is possible that this proportion has increased since. Obesity is a major risk factor for cancer. The malignancies associated with obesity include cancers of the oesophagus, colon and rectum, breast, prostate, liver, gallbladder, pancreas, and kidney.[56] In a sixteen-year prospective study of almost 1 million US adults (free of cancer at enrolment in 1982), significant obesity increased the risk of cancer in women by 62 per cent, and in men by 52 per cent.[57] The main mechanisms proposed to explain this risk include metabolic and hormonal derangements, and the greater propensity for inflammation and oxidative stress,[58] a significant degree of which is programmed in early life and adolescence (Chapter 4).

Specific dietary factors have been harder to investigate. Based on some studies, diets high in saturated fat and red meat but low in fresh fruits, vegetables and fibre are general risk factors for cancer. But findings have been inconsistent.[59-61] The role of processed meats or red meat cooked at high temperatures has been implied but not confirmed in human studies. Several dietary factors are strong risk factors for specific cancers, including chewing betel quid with oral cancer[62, 63] and aflatoxin B1 (from a fungus contaminating grains and nuts) with liver cancer, seen commonly in developing countries.[64]

Most health agencies advocate healthy dietary patterns and weight control to reduce the risk of cancer and improve the survival of affected people. This is largely common sense. There is also an extensive range of naturally occurring elements – found in fruits, vegetables, teas, spices and wholegrain cereals – that has many and various anti-inflammatory and antioxidant properties. Many of these 'chemopreventive compounds' (such as carotenoids, vitamins, resveratrol, quercetin, silymarin, sulphoraphane and indole-3-carbinol, and Thymoquinone) have been shown to have favourable effects in animals.[65] Some of these experimental studies are quite extensive and convincing, but data in humans is more limited. The general advantage of these compounds is that they are safe and usually target multiple cell-signalling pathways.

Particular attention to healthy nutrition in pregnancy and childhood remains the most effective ways to promote healthy patterns of early weight gain and to reduce the risk of adult obesity – and to establish healthy metabolic responses and healthy eating behaviours (Chapters 4 and 5).

Physical inactivity and a sedentary lifestyle are major risk factors for many chronic diseases (Chapter 8), including cancer.[66] The protective effects of physical activity have been best shown for cancers such as colon cancer,[67] breast cancer,[68] and for uterine cancers.[69] For other cancers (prostate, lung, ovary, pancreas and stomach cancers) there is also suggestive evidence.[70] Just as inactivity increases risk through multiple pathways, so too are there many protective effects of physical activity, including through hormones, improved immune function, and reduced inflammation.

There are other environmental contaminants implicated in specific cancers including asbestos in mesothelioma, and benzene in leukaemias. But these factors are usually also modulated by other physiological processes or environmental exposures. High levels of heavy-metals pollution, particularly in developing regions, have also been linked with increased risk of many cancers through food contamination.[71, 72] Life-long exposure to these contaminants has implications for many aspects of health. These and many other environmental factors (diesel exhaust particles, polycyclic aromatic hydrocarbons from fossil fuels, fungi, and other indoor and outdoor pollutants) may have both mutagenic and epigenetic effects that promote cancer risk.

Infections may contribute to around 10 per cent of cancers in developed countries, but as many as 25 per cent of cancers in developing regions.[73] *Helicobacter pylori* in the stomach is a high risk factor for stomach cancer and can induce epigenetic changes (DNA hypermethylation) in cells lining the stomach, consistent with the aberrant patterns seen in gastric cancer.[74] Inactivation of important tumour-suppressor genes can contribute to malignant transformation.[75]

Several chronic viral infections have been associated with cancer. The association between viral hepatitis and liver cancer has been recognised for many years, but newer studies have revealed that infection with the hepatitis B virus is associated with early epigenetic events and progressive changes leading to cancer.[76] Similarly, the epigenetic changes induced by HPV infection in the cervix could ultimately provide an avenue to gauge the level of precancerous transformation.[77] Treatments to eradicate *Helicobacter pylori* and vaccinations against hepatitis and HPV have done much to reduce the risk of cancer from these agents. One of the best recent success stories is the introduction of the HPV vaccine (for human papilloma virus) to prevent cancer of the cervix.[78]

But there are other cancer-associated viruses for which there are, as yet, no vaccines available. Notorious in this regard is the Epstein Barr virus (EBV) that infects more than 90 per cent of humans. Those who contract EBV for the first time as adolescents or adults may develop symptoms of 'glandular fever'. EBV persists in the cells it infects – often the immune cells (B cells) and the cells (epithelia) lining our nose and throat – usually as a 'harmless passenger', for the rest of our lives. It circularises its DNA and hides in the nucleus of our cells where it can evade detection by the immune system. Under certain conditions the virus can reactivate and switch on genes that transform and immortalise the cells it resides in – triggered, perhaps, by unrelated infections or environmental exposures. This transformation can lead to a number of human cancers, particularly those affecting bone marrow, lymph and the immune system (lymphomas). In regions of Asia and Africa, EBV is also associated with very high rates of nasopharyngeal (nose and throat) cancers. Some of the EBV genes responsible for malignant transformation include miRNAs (above) produced by the virus itself which interfere with the cells' normal defence and regulatory systems. If a safe and effective vaccination can be eventually developed, this will have a significant impact on a range of cancers.

Radiation, both natural and man-made, is another important risk factor for cancer. Its effects only became obvious decades after the discovery of radioisotopes and x-rays at the end of the nineteeth century. In the meantime, x-rays and fluoroscopes had been used widely as side-show attractions, exposing many to needless radiation. X-ray fluoroscopes were installed in shoe shops as shoe-fitting machines. Women working in factories painting clock dials began developing rare cancers from licking the radium on their paintbrushes. But until the bombings of Hiroshima and Nagasaki, the effects of high-dose radiation exposure were not fully appreciated. Marie Curie, who discovered radium, warned of the possible health effects, and later died of bone-marrow failure, the likely effect of radiation exposure.

Today, radiation – including ultraviolet (UV) radiation, particularly UVB – may contribute to around 10 per cent of malignancies.[79] Malignant and premalignant skin lesions are extremely common in countries of high sun exposure, such as Australia. Skin cancers are common in old age, with initiating events probably taking place in childhood. Single, brief exposures to higher radiation doses can increase the risk of cancer, often manifesting many years later. Radiation from modern x-ray procedures is well below the range absorbed from the natural environment, but whole-body CT scans deliver much higher radiation doses, and may contribute to a small proportion of cancer cases. It is important to limit unnecessary imaging, particularly in pregnancy. Exposure to radiation in childhood carries a much higher of cancers (such as leukaemia) compared with exposure in adulthood; and this risk is *magnified* even further with radiation exposure *before* birth. The effects of other sources of low-dose radiation – such as frequent air travel, non-ionizing radio frequency radiation from mobile phones, and electric power transmission – cause public concern, but we still know very little about how these might be affecting our health.

Radiation penetrates deep into cells and has unpredictable effects. If it hits a chromosome it can cause major structural damage. In the extreme, this will kill cells. But smaller degrees of damage can permit survival of some cells with damaged genes, causing a greater predisposition for malignant transformation, especially if tumour-suppressor genes are damaged.

FIGHTING CANCER ON ALL FRONTS: EACH PERSONAL BATTLE AND THE LONG GAME

At the moment, most of our investment into cancer is in fighting each important battle in each person who develops the disease. Developing new technologies and new, more-effective treatments must remain a research priority. But we have to fight the war on a wider front at the same time. We need to reduce the global risk of cancer across the population.

We are learning that many cancers are the result of accumulated epigenetic modifications – with specific or general effects – induced by multiple environmental exposures, accumulated over various lengths of time, and appearing at different life stages.[80] Because of the cumulative, extended process, pinpointing specific causes will remain impossible for the foreseeable future. The best we can do is to minimise the things we know increase our risk, and do this as early as possible. To be most effective, this must be done at both the individual level and the societal level.

In many cases cancer has the very same risk factors as other common NCDs; so the benefits of modifying these lifestyle behaviours and environmental risks would be manifold. Not all of this 'risk' is under our individual control, but without taking greater responsibility at both the individual and societal level, it will be difficult to overcome the growing burden of cancer and chronic diseases.

10

Asthma, allergy and immune diseases

In 2011, after many long months of writing, I published my first book, *The Allergy Epidemic, A Mystery of Modern Life,* to help raise public awareness of how common and how serious allergic diseases can be.[1] Like many other medical scientists, I had spent long hours researching the causes of this modern epidemic and writing papers to my professional colleagues debating and exploring different concepts.

After writing that book – with a particular focus on 'early-life origins' of allergy – there's no need to repeat the process here. But I will pull out some bits and pieces that help this broader 'DOHaD' story, including new developments in our understanding of how this epidemic links with the rise of other NCDs.

A THOROUGHLY MODERN EPIDEMIC:
Once associated with 'affluence', allergic diseases are now rapidly rising in developing regions, and it is estimated that around 30 to 40 per cent of the world's population are affected by allergic disease.

Although both ancient and historical texts describe what we might infer to be cases of allergy, these appear to have been fairly uncommon. 'Hay fever' was first described in modern medical literature in 1819, and in the ten years that followed, only twenty-eight similar cases had been found.[2] Today, by contrast, around

40 per cent of populations in high-income countries experience hay fever (allergic rhinitis) at some stage.[3]

Allergy and asthma are fast becoming universal. Australia has one of the highest rates of allergic diseases, reflecting similar trends in other countries of Oceania, Western Europe and North America over the last fifty years. The same patterns are now emerging in regions of the world undergoing economic transition and Westernisation, including Asia, Africa and South America. The emergence of allergy in remote indigenous islanders of Papua New Guinea underscores that this has become a truly global epidemic.

Allergic disease includes eczema, food allergy and anaphylaxis, asthma, and allergic rhinitis – all driven by inflammation and inappropriate immune responses to the environment. Food allergies and eczema appear first, with more than 25 per cent of infants developing eczema and 20 per cent already becoming sensitised to common foods such as egg, milk and peanuts in the first year of life. Asthma becomes more common with age, affecting around 25 per cent of school-aged children. By adulthood around 40 per cent of the population in high-income countries are sensitised (with positive allergy tests) to common inhaled allergens such as dust mites, pets and pollens (reviewed in detail in *The Allergy Epidemic*).

WHAT GOES WRONG IN ALLERGIC DISEASE?

The immune system is an incredible network of cells and molecules that are present in almost every tissue and every organ of the body; all with different functions, all constantly interacting at lightning speed. With every interaction with the environment the immune system gathers its own 'memories' about *if* it needs to 'react' and what *kind* of reaction might be best. It uses these for rapid responses if the same 'threat' is ever encountered again.

On the frontline there are a number of cells that are 'hardwired' to defend against threats. These 'innate' immune cells provide some basic protection without any need for programming. These cells call for help from more specialised defence teams of 'naive' lymphocytes, which produce more-efficient adaptive responses following 'training' by the innate cells.

Innate cells digest new encounters into small pieces (called 'antigens') and 'present' these to the immature lymphocytes to teach them about the environment and to program their responses when they encounter the same antigen again. Innate 'antigen presenting cells' (APC) called macrophages also secrete signals (called cytokines) that influence the pattern of lymphocyte response.

The main lymphocytes in our specialised defence system are the B cells and the T cells (also called 'effector' cells). They mount our customised 'adaptive' immune responses and orchestrate the type of attack action. B cells make antibodies to neutralise a specific protein (antigen). Some T cells can destroy the antigen they have been trained to recognise, others produce cytokine signals to tell B cells what kind of antibodies to make (including allergic and normal defence antibodies).

The adaptive immune system allows symbiotic relationships with the microbial world, selectively promoting beneficial microbes for metabolic and physiological gain.[4] Quicker, more-targeted and selective, adaptive immune responses also ensure more efficient use of energy.[5]

Once a threat has passed, the immune reaction must be switched off quickly, or we would soon die from uncontrolled inflammation. There are a number of mechanisms that do this, including 'regulatory cells' and 'regulatory cytokines' which have inhibitory effects on other immune cells to reduce collateral damage. These complex regulatory systems are also strongly dependent on microbial signals, particularly in the gut.

So what goes wrong with this system in allergy? Allergy and autoimmunity occur when adaptive responses are targeted inappropriately. At some point after innate cells have presented an antigen to the T cells, the T cells 'decide' to produce Type 2 allergic cytokines instead of normal Type 1 defence responses. The Type 2 cytokines then induce the B cells to make allergic 'IgE' antibodies, which are responsible for all of the signs and symptoms of allergy. Inadequate inhibitory signals from regulatory cells may also play a role in allowing an allergic response to become more established.

Bacterial exposure stimulates innate cells and promotes the Type 1 responses because these innate responses are inherent for microbial defence. These innate responses also inhibit Type 2 allergic responses. Bacteria also promote activation of regulatory cells. This is why it has been proposed that cleaner environments could be promoting allergic disease by sending inadequate signals to the Type 1 and regulatory cells, allowing the Type 2 allergic responses to expand unchecked.

YOUNG CHILDREN BEAR THE BRUNT OF THE ALLERGY EPIDEMIC:
The latest generation of young infants bears the brunt of an epidemic of potentially life-threatening food allergies which were uncommon in their parents and rare in their grandparents. In just ten years we have seen a five-fold rise in serious (anaphylactic) food allergies in preschoolers.[6] Professor Katie Allen's 'Healthnuts' study of over 5,000 Melbourne infants revealed that more than 10 per cent of one year olds now have a food allergy. Even more have other symptoms, such as eczema.[7] This is the highest incidence ever reported anywhere in the world.

Just when we were feeling confident that asthma had reached a peak in the industrialised countries, this 'second wave' epidemic of food allergy began to appear. We are also seeing more-severe disease, earlier onset and delayed resolution. Common food allergies, such as egg and milk allergy, were previously transient in early childhood but are becoming more persistent.

It is still quite baffling why this surge in food allergy has lagged so far behind the epidemic of asthma and other allergies.

We are not sure the same environmental changes are responsible, or if something new is happening. More research is needed, but allergic diseases are still not considered to be a funding priority by many governments.

POWERFUL EFFECTS OF THE EARLY ENVIRONMENT ON IMMUNE HEALTH:

As with other NCDs, genetic factors influence susceptibility but cannot explain the surge in allergic disease in recent generations, and environmental factors must play a role. Other immune diseases have had similar increases during the same period. Cleaner environments and declining exposure to infectious agents have been one of the leading explanations for this dual increase, but there are likely to be a number of other factors including modern diets and pollutants.

Unlike allergy, where the 'effector' immune responses are directed against environmental antigens, in autoimmune diseases the inappropriate immune responses are directed to 'self' antigens (Chapter 5), leading to a different kind of inflammation and self-destruction. But although the 'target' of the misdirected 'effector' T cell response may be different, the underlying tendency for immune 'dysregulation' is a common element in all of these conditions.

This has turned attention to the immune pathways that 'regulate' our immune system (below). 'Regulatory' immune responses are critical in 'switching off' normal defensive 'effector' responses once the danger has passed. They are particularly important in suppressing or destroying effector cells that do harm, for example targeting self-tissues or harmless environmental proteins, such as foods. These 'regulatory' pathways develop much earlier in life than previously suspected, and are strongly influenced by the early environment, as we will see in a moment.

Allergic diseases are arguably are the *only* common NCDs to appear in the first year of life, leaving no doubt about the early vulnerability of the developing immune system. As a measure of immune health they are, therefore, a useful *early barometer* of environmental impact, and an early measure of effectiveness of any interventions that we might try to prevent disease. This is why we need to make sure allergy eventually finds a place in the global NCD agenda.

NEW INSIGHTS INTO WHY ALLERGIC RESPONSES EVOLVED:

Type 2 allergic responses are *not* inherently pathological and only cause disease when excessive or misdirected. The very rapid 'acute' IgE-mediated responses appear to have evolved to *protect* against a broad and diverse range of environmental irritants, toxins, venoms, parasites (such as mosquitoes and ticks) and noxious environmental chemicals (xenobiotics) including naturally occurring chemical compounds in plants (phytochemicals).[8] It allows us to sense these threats at *very low* levels in the environment, and react with rapid histamine reactions induced by IgE antibodies: sneezing, tearing, coughing, vomiting, diarrhoea and itch, which expel or remove these threats – helping our ancient ancestors survive. Moreover, these unpleasant symptoms also change our *behaviour*, conditioning us to avoid that noxious thing in the future.

This explains both the *symptoms* of the allergic response, the *urgency* of response, and the types of *allergen triggers*. Most 'modern' allergens are chemically related to the ancient environmental threats the Type 2 responses evolved to react to, including plants (pollens and food allergens), mites, venoms, antibiotics (from moulds), animals and fungi. This does not explain, however, the dramatic increase in the tendency for misdirected allergic responses in the last few decades.

The answer appears to lie with the lifestyle changes we have experienced in a relatively short period of time, and how our evolutionary training may lead to maladaptive responses under these modern conditions.

Evolutionary adaptation to an over-abundant environment – close links between immunity, body weight and inflammation:

We have already considered (Chapter 4) how we have come to show all of the hallmarks of animals living in captivity. We are prone to obesity and we are prone to inflammation. It is now apparent that these changes in our metabolic responses and immune responses may in fact be linked and, once again, these are driven by evolutionary mechanisms.

In times of scarcity, organisms live longer if they conserve energy and 'tolerate' non-lethal threats. But in more favourable conditions there is greater benefit in mounting a strong immune response. We can see these dramatic shifts in metabolic and immune function in animals that undergo significant environmental changes. The Siberian hamster, like many seasonally breeding rodents, has evolved complex adaptations to maximise survival and reproductive success. To cope with winter's scarcity and cold, these small animals reduce energy-demanding activities that are not essential for immediate survival. Mounting a strong immune response requires energy. Under winter-like conditions animals not only reduce their body mass and adipose tissue mass, but they also show significant changes in immune function, reducing their inflammatory responses. In summer-like conditions Siberian hamsters have significantly higher levels of circulating inflammatory cytokines (such as interleukin-6) and produce higher fevers.[9] Their 'innate' antigen presenting cells (APC) also show significantly increased activity in abundant conditions.[10] These findings are consistent with the immune suppression seen in human malnutrition, and the increased inflammatory responses seen in obesity.

It is no coincidence that we are increasingly prone to *both* immune diseases and obesity-associated NCDs. But it is surprising that we have been so slow to realise that these are directly linked.

We now know there are complex, intricate relationships between metabolism and immune function.[11] The same hormones that regulate appetite, fat storage and metabolism also regulate immune function. Leptin is one of the main hormones involved in fat metabolism, and, as we have seen earlier (Chapter 4), much higher levels (around five times) are seen in obesity. Under conditions of food deprivation and reduced body-fat mass, lower leptin levels reduce metabolic expenditure to conserve energy.[12] In the Siberian hamsters it is actually the lower leptin levels that suppress *inflammatory* responses during winter – an effect that can be reversed artificially by infusing leptin.

Leptin also enhances both innate and adaptive immune responses, and is part of the family of inflammatory cytokine (the interleukin-6 family). It increases the ability of innate immune cells to engulf and kill bacteria. It also stimulates adaptive immune responses by inducing proliferation or cytokine secretion by T cells. Leptin is linked to an increased risk of both allergy[13] and autoimmunity.[14, 15] Higher leptin levels are associated with other 'risk factors', including reduced psychological stress[16] and physical inactivity,[17, 18] independent of body weight.

Insulin is another hormone important for both energy balance and immune function. As seen in Chapter 5, insulin enhances uptake of nutrients, increasing cellular metabolism, energy requirements and protein synthesis. All of these factors also promote lymphocytes' activity and T-cell responsiveness. Under conditions of high blood sugar, increased insulin secretion promotes the production of inflammatory cytokines and T-cell activation.

As fatty (adipose) tissue accumulates it expresses and releases increasing levels of inflammatory cytokines such as interleukin-6, inducing low-grade systemic inflammation[19] (elevated CRP; see Chapter 4). In early childhood this is associated with allergic sensitisation, in particular to foods. Allergic IgE antibodies are higher among overweight children, although the exact nature of this relationship is still being explored.[20]

This nexus between metabolism and the immune system clearly occurs at many levels: hormonal interactions, sensing of nutrients by immune cells, and in the gut bacteria (microbiota).

Changes in modern nutrition are contributing to obesity and both are implicated in changes in immune health. The chronic inflammatory states of obesity, diabetes, atherosclerosis and other NCDs appears to be a chronic tissue *malfunction* and a shift of the normal homeostasis or 'balance' to adapt to new physiological or metabolic conditions.[21] A significant component of this risk for inflammation is programmed early in life (Chapter 6).[22]

THE EPIDEMICS OF ALLERGY AND OBESITY: A TWO-WAY STREET?

Medicine is predisposed to operating in specialities and silos, and so the connections between allergic diseases and other NCDs have not been extensively recognised or explored. But some researchers have started to investigate these.

Several studies suggest that childhood obesity increases the risk of asthma[23, 24] and food allergy. There has been interest in how the metabolic changes in obesity may contribute to the risk of allergy and inflammation in the airways. But it must also be remembered that the effects of the obesity epidemic actually begin before birth.

Maternal obesity in pregnancy is a source of chronic low-grade inflammation for the developing foetus, with higher levels of inflammatory cytokines in the circulation[25] and in the placenta.[26] There is some preliminary evidence that children of overweight mothers have an increased risk of asthma and lung disease[27, 28] and that the levels of adipokines in cord blood may influence the risk of wheezing.[29] Again, more research is needed here.

Once allergy has developed, obesity may exacerbate the symptoms of asthma and other allergic diseases. In sensitised animals higher leptin levels, as seen in obesity, induce exaggerated allergic IgE antibody responses and increased airways' inflammation.

However, other studies suggest a more complex interplay. Leptin levels are higher in allergic than nonallergic individuals, even after allowing for differences in body weight.[30] This suggests that allergic inflammation may have additional effects on leptin levels and, consequently, metabolism; in other words the relationship between allergy and obesity might operate in *both* directions. Animal studies support this, showing serum leptin is increased during allergic reactions in the airways. Allergen exposure not only induces inflammation in the airways, but also caused changes in the composition of adipose tissue and the levels of other adipokines in the blood stream. This is consistent with what is seen in human allergy. Leptin levels are very much higher in pollen-allergic individuals when exposed to pollen.[31] So, leptin may increase the systemic inflammatory response resulting from allergen exposure, which in turn causes a greater release of leptin from fat stores.

Some of the prime suspects in the allergy epidemic are now emerging as likely culprits in the obesity epidemic. The 'hygiene hypothesis' has long been a major contender in the rise of allergy, with cleaner environments reducing the microbial diversity of our intestinal bacteria. Our modern (low fibre, high fat) dietary patterns add to this reduced biodiversity and allergy risk. For these reasons, prebiotic fibre and probiotic bacteria have both been used to prevent allergy in early life (below). These potential treatments and prevention strategies are now being considered in preventing obesity (Chapter 4) to address the metabolic effects of the same risk factors.

There are also emerging links between other immune diseases and obesity that may be mediated through changes in levels of adipokines such as leptin.[32] In animals, a role of leptin in auto-immune diseases has been suggested by the fact that 'leptin-deficient' mice are resistant to a range of autoimmune diseases such as autoimmune arthritis, autoimmune liver and bowel disease, and autoimmune encephalomyelitis – an animal equivalent of multiple sclerosis. When leptin is artificially administered to animals it can trigger the development of autoimmunity, as well as accelerate the progression of disease. In humans, more studies are needed to determine the potential influence of leptin on autoimmune disease.

ALLERGY AS A RISK FACTOR FOR CARDIOVASCULAR DISEASE:

It is not widely known that allergy increases the risk of heart disease and other NCDs. People with asthma, allergy or elevated IgE have increased carotid atherosclerosis[33, 34] or strokes.[35, 36] In fact, men with common allergic diseases have almost *four times* the risk of developing atherosclerosis.[37] Day-to-day variation in pollen concentrations have also been linked to deaths due to cardiovascular disease,[38] suggesting in allergic individuals this may tip a delicate balance in those with atherosclerosis. Treatments that control the level of inflammation in asthma reduce the risk of cardiovascular disease, although this is not seen in all studies. Animal studies lend further support to this association, with increased cardiovascular changes seen in experimentally induced allergy and asthma.

Given that we now understand cardiovascular disease as an inflammatory condition of blood vessels, this should be no surprise. The 'foamy macrophages' that accumulate cholesterol and drive the inflammation of atherosclerosis are 'antigen presenting cells' (APC) – the very same cells that program allergic responses. Macrophages are sensitive to the conditions in the surrounding tissues; in our blood vessels they can become activated by inflammatory cytokines from other parts of the body. Allergic inflammation in various tissues can do just that.

Allergy is a truly 'systemic' condition. Although the symptoms may *appear* localised to certain tissues (such as the skin, the nose, the chest or the gut) there are much more *generalised* changes in the immune system. Let's take the example of allergic rhinitis (hay fever) and the related greater risk of heart disease.

During an allergic reaction in the nose, there is a surge of cytokine released into the blood stream, designed to recruit inflammatory cells into the local tissues to defend against the perceived 'threat'. This triggers the bone marrow to release showers of stem cells (immune-cell progenitors) into the circulation so that they can follow homing signals to the site of inflammation, essentially adding fuel to the fire. This inflammation associated with common allergic diseases may explain the higher risk of atherosclerosis seen in these individuals.

With the new generation of allergy-prone infants we might do well to anticipate, and hopefully prevent, a surge in cardiovascular risk.

EFFECTS OF ALLERGY ON OUR BRAIN AND BEHAVIOUR:
It is called hay 'fever' for good reason. People who suffer badly from allergic rhinitis feel tired, weak and generally unwell – as though they have a fever or the 'flu. As with viral infections, cytokines are released into the bloodstream from the inflamed tissues. It is actually our immune *response* that makes us 'feel sick'. Cytokines reach our brain to influence our mood and behaviour.[39] They can also cause fever and other symptoms. We become less active, sleepier, and less sociable, helping us rest and fight the 'threat'. People with chronic allergic rhinitis experience many of these 'sickness' symptoms (as do people with depression, see Chapter 7).

In adults, allergic disease has been shown to impair performance and cognitive abilities and to increase symptoms of anxiety and depression. Allergy in children, particularly allergic rhinitis, can affect school performance and many are misdiagnosed with attention deficit hyperactivity disorder (ADHD). In our own studies we have shown that even in the first years of life, allergic disease is associated with differences in neurodevelopment and behaviour.

In the extreme, allergic disease has been linked to suicide risk. Suicide victims have elevated levels of inflammatory cytokines (including interleukin-6) in their brain tissue and their circulation.[40] They also show dysregulation of their stress responses ('HPA axis' function; Chapter 7).[41] Although it is not possible to prove a 'causal' association, studies of almost 38,000 suicide cases between 1970 and 2001 have shown peaks in suicide during the pollen season in Europe.[42] This has evoked speculation that the systemic inflammation of allergy may precipitate or exacerbate depressive episodes in vulnerable people. One study also reported higher levels of Type 2 allergic cytokines in the brains of suicide victims.[43] The same researchers have shown that when rats and mice are sensitised and exposed to allergens they show increased aggression,[44] anxiety-like behaviour and reduced social interaction; this was associated with an increase in Type 2 allergic cytokines in their brains.[45] A lot more research is needed in this area to understand these associations.

Preschool children with higher cortisol levels (associated with stress response; Chapter 7) also appear to be at greater risk of allergic sensitisation and allergic symptoms,[46] suggesting that very early alterations in stress responses may contribute to the development of allergic disease. We and others have also shown that cellular immune function (including the capacity and pattern of cytokine production) is influenced by other antenatal factors (including maternal smoking and dietary factors) as well as intra-uterine infections. Many of these factors have been implicated in both allergic risk and neurodevelopmental problems.

We have long suspected a link between the nervous system and the immune system (Chapter 7). Although we still don't understand these connections well, we can see that the interaction between our brain and our immune system is likely to be complex and 'bi-directional'.[47] Our emotional responses play a role in exacerbating allergic reactions; our immune reactions affect our state of emotional well-being. These intricate 'neuro-immune' interactions begin very early in development and have significant capacity to shape the subsequent function of many systems.

ENVIRONMENTAL EFFECTS ON IMMUNE PROGRAMMING BEGIN IN FOETAL LIFE:

It was once believed that the foetus did not react to its environment, but we now know that it can mount immune reactions of its own. And the reason that it *appears* to tolerate its environment so well is because it has developed sophisticated regulatory mechanisms to actively control its own immune responses. It learns to recognise and tolerate its own tissues in this early period. The specialised regulatory immune cells (called regulatory T cells or 'T-regs') that do this are found in much higher numbers in early foetal life than previously realised. In fact, the foetus appears to be actively pre-disposed to generate 'regulatory' rather than 'effector' responses.[48]

These early regulatory responses are strongly influenced by the early environment. Early microbial exposure is a critical driver for the optimal development of these regulatory pathways. It has become clear that naturally resident friendly (nonpathogenic) bacteria such as lactobacilli and bifidobacteria in the gut are probably even more important in inducing regulatory T cells (T-reg). The well-balanced relationships between our immune system and our nonpathogenic inhabitants have mutual benefits. Our immune systems tolerate these bacteria and selectively allow them to thrive. On the other hand, the microbes suppress pathogens, and encourage and maintain our regulatory responses. This constrains our potentially pro-inflammatory effector T-cell clones, and curbs the risk of allergy, autoimmunity and other NCDs.

During pregnancy there are enormous opportunities for the maternal environment to influence foetal immune development.

In our own studies we have shown that a range of maternal exposures in pregnancy are associated with changes in the patterns of foetal immune responses, including microbial exposure, dietary nutrients and environmental pollutants.[49] It is possible, and likely, that these factors also contribute to the emergent differences in immune function that are already evident at birth in children who go on to develop allergic disease in the first years of life.[50]

TYPE 2 IMMUNE RESPONSES ARE NORMAL IN PREGNANCY AND THE PERINATAL PERIOD:

Complex immunologic mechanisms have evolved in the placenta to allow the foetal and maternal immune systems to coexist. Pregnancy hormones influence the maternal immune system to adapt to a more 'Type 2 state' in order to down-regulate Type 1– mediated tissue rejection of 'foreign' foetal antigens.[51] Pregnancy hormones, such as human chorionic gonadotrophin (hCG), also attract regulatory T cells (T-reg) to the materno–foetal interface. These cells also play a critical role in maintaining immune tolerance between the mother and the foetus for successful pregnancy.

Contrary to traditional expectations, human foetal T cells are responsive as early as twenty-two weeks' gestation. Foetal immune responses are also sensitive to the ambient cytokine environment, and reflect the 'Type 2 skewed' immune milieu of pregnancy. This 'allergic' pattern of immune response is *normal* in the newborn period. Our own landmark studies established that Type 2 dominant responses of the newborn must undergo maturation towards more mature Type 1 dominant responses in the early postnatal period.[52] Allergy appears to result from the failure of this normal maturation process and a persistence of the less mature Type 2 responses.

We also showed evidence of *presymptomatic* immaturity in newborns who later develop allergic disease, confirming that 'the scene is set' to some extent before birth. Although this affects several aspects of neonatal immunity, less-mature Type 1 and regulatory T cell (T-reg) function are particularly implicated, by increasing the risk of persistent Type 2 immune responses to allergens. Notably, our work emphasised the likely importance of microbial exposure as an important factor in the maturation of Type 1 and T-reg responses. But we now know that the level of microbial exposure *before* birth may also influence foetal immune maturation and the risk of allergic disease.[53]

THE WOMB IS NOT STERILE AFTER ALL:

The long-standing assumption that the womb is 'sterile' is also turning out not to be true. More-sophisticated modern technology has revealed whole new ecosystems in unexpected places. And some species of bacteria are an important part of healthy pregnancy. 'Friendly' bacteria, sometimes called 'symbionts', colonise and help protect our skin, intestines, airways and all other parts of our body surfaces, inside and out.

Evidence suggests that transfer of maternal microbiota begins during pregnancy, providing the foetus with a 'pioneer' micro-biome.[54] In normal pregnancies, microbes can be detected in amniotic fluid, placental and foetal membranes, cord blood and meconium. Animal studies have shown tagged bacteria being transferred from mother to foetus during normal pregnancy.

The microbes we receive from our mother may provide an initial source of stimulation for immune development. They may also help prepare us for the much larger bacterial inoculum we receive during vaginal delivery and breastfeeding. Antenatal exposure may serve to increase the chance for optimal 'mutualism' with our bacteria after birth.

So far, we have an inkling that the profile of colonisation associated with high-fat, low-fibre diets, obesity and metabolic dysregulation is 'unhealthy', but we have not yet defined the range or scope of what might be considered 'ideal' colonisation.

My colleague in Germany, Professor Harald Renz, and his team, have found that giving pregnant animals bacterial products, symbiotic bacteria such as lactobacillus, or other harmless bacteria from the environment, reduced or prevented allergic disease in the offspring. These effects appeared to be mediated by activation of maternal innate immune pathways (toll-like receptors), with some associated epigenetic changes in Type 1 immune genes and the microbial responses of the newborn offspring.[55]

When we look to humans, there is also evidence that mothers who live in 'high microbial environments' during pregnancy are less likely to have allergic children.

Bavarian farmhouses are built with family dwelling space in the same structure as the barn and cattle stalls. The levels of bacteria and bacterial products such as endotoxin are substantially higher than in non-farming houses.[56, 57] Children growing up in this setting have much less allergy, hay fever, and asthma.[58] And *antenatal* exposure affords greater protection from allergy than postnatal exposure alone.[59] At birth they have increased numbers and function of regulatory cells (T-reg), lower Type 2 immune responses and higher Type 1 responses, likely to be important for suppressing the development of allergy after birth.

We still know little about how this evolves in pregnancy, however, and we need to further explore the relationship between maternal gut flora in pregnancy and immune development.

To date, most strategies to prevent allergic disease with bacterial products have started prebiotic fibre and/or probiotics in the postnatal period or only very *late* pregnancy (typically the last two to four weeks), with the aim of enhancing colonisation immediately after birth.[60] These studies have shown some protection from eczema, although there has been no consistent reduction in allergen sensitisation, asthma or other allergic disorders.[61]

Even using probiotics in the last two weeks of pregnancy has shown significant immune effects on the foetus in some studies

– including altered expression of innate immune genes in the placenta and in the foetal gut, and, in our own studies, some increases in cord blood Type 1 cytokines.

It may be more logical to investigate the effects of probiotics (and prebiotics – discussed further below) much earlier in pregnancy, at a time when foetal metabolic and immune responses are first initiated. A recent large-scale observational study of 40,614 Norwegian mother–child pairs found that probiotic milk consumption in pregnancy (assessed at twenty-two weeks gestation) was associated with a reduced incidence of atopic eczema and allergic rhinitis at three years of age.[62] We now hope to conduct trials using prebiotics and probiotics much earlier in pregnancy to see if changes in maternal gut flora can improve both immune and metabolic programming of the foetus to reduce the burden of not only allergy but other common inflammatory diseases.

OTHER MATERNAL FACTORS IN PREGNANCY:

Many of the factors that have been implicated in the allergy epidemic have been discussed at length in my first book. This is only intended to be a brief summary.

a) maternal diet:

Maternal diet has recognised effects on immune development and allergic risk. This includes both specific nutrients and common dietary patterns. The Mediterranean diet appears to protect against wheezing in early childhood, although this has not been seen in all studies. Our own studies have shown that fish-oil supplementation from twenty weeks' gestation had effects on immune function, with reduced inflammation and oxidative stress in the newborns at risk of allergic disease because of an immediate family history. We also saw a reduction in egg sensitisation and eczema severity at twelve months of age.[63] Postnatal fish-oil supplementation appears to be less effective than improving n-3 PUFA levels in pregnancy.

The importance of optimising folic-acid status in early pregnancy for neural-tube defects is undisputed. However, *after* this risk period there are emerging associations with allergic disease in the offspring, including atopic dermatitis[64, 65] childhood wheeze[66] and asthma.[67] In our own human studies, we have shown that variations in folate levels in maternal and cord blood are associated with epigenetic changes in immune cells of the newborn. More studies are needed before any changes are made to current recommendations.

Other nutrients of specific interest in immune health include vitamin D, antioxidants (such as vitamins A, C, and E) and prebiotics (discussed further below). Of these, only vitamin D (Chapter 8) is currently under investigation in randomised controlled trials in pregnant women, and the specific role in allergy prevention is still unclear.

Many women report considerable confusion, fear and guilt about their diets in pregnancy. Professional advice is lacking or conflicting, and many source information for themselves. Given just how important diet is for so many aspects of development, the lack of clear and consistent advice for pregnant women must be regarded as a failing that needs to be addressed urgently.

b) Maternal smoking and other environmental pollutants:
The toxic effects of maternal smoking on placental function and foetal development are considerable. This includes specific effects on lung growth and asthma risk.

We also now know that the oxidative stress produced by cigarette smoke and air pollution can have significant epigenetic effects including remodelling of pro-inflammatory genes. Traffic exhaust particles can also have epigenetic effects in pregnancy. Mice exposed to diesel-exhaust particles show increased production of IgE allergic antibodies.[68] Extending this to human pregnancy, another team discovered that high levels of maternal exposure to traffic particles were also associated with epigenetic changes and increased risk of asthma in the children.[69]

New chemicals and pollutants include a range of organic products of industry and agriculture, including polychlorinated biphenyl compounds (PCBs), organochlorine pesticides, dioxins and phthalates. These contaminate modern homes, food, clothing, and water sources, accumulating in human tissue with age. They can have toxic effects at high doses, including on the immune system,[70] but the effects of chronic low-level exposure are less well understood. There is some evidence that low-level contamination may more selectively inhibit Type 1 immune responses and favour allergic Type 2 immune responses through their 'oestrogenic' hormonal activity. Some of these 'hormone imposters' have been readily measured in breastmilk, cord blood and placental tissue, highlighting the potential to influence early development. In our own studies we detected organic pollutants (particularly organo-chlorine pesticides) in more than 90 per cent of abdominal fat samples from mothers undergoing caesarean section and more than 60 per cent of breastmilk samples.[71]

It is difficult for individuals to avoid these many ubiquitous contaminants, and an important responsibility of governments and industry to work to further reduce them.

c) Maternal stress in pregnancy:

We have already seen (Chapter 7) how our hypothalamic-pituitary-adrenal (HPA) axis stress responses and immune system are closely linked. The production of cortisol (in response to stress) by the adrenal gland modulates immune responses, and in pregnancy this can influence the placental immune system. Stress, inflammation and activation of the maternal 'HPA axis' induces placental Type 1 immune-response cytokines, which are harmful to the foetus in high levels.

d) Effects of delivery method and maternal medications in pregnancy:

Medical interventions have significantly reduced maternal and infant mortality over the last century. However, overuse of medications that are deemed safe in pregnancy, such as antacids, reflux medications, paracetamol, and antibiotics, may carry an increased risk of some diseases, such as allergy.

A flurry of publications has reported the use of paracetamol in pregnancy as a risk factor for childhood asthma.[72] The depletion of antioxidants by paracetamol may predispose to both abnormal lung function and immune function, and although these effects are not large,[73-76] this risk is avoidable and requires further investigation.

Acid-suppressive medications in pregnancy are also associated with an increased risk of childhood allergy.[77] It has been proposed that these medications may interfere with the normal denaturation of ingested allergens in the stomach. They may also affect bacteria in the stomach, such as *Helicobacter* (below).

Antibiotics may affect microbial colonisation patterns in pregnancy, increasing the risk of asthma and eczema and food allergy. Maternal antibiotics prior to or during delivery reduce the transmission of healthy, normal vaginal lactobacillus flora to the neonate at birth. Clearly, antibiotics should be used when the immediate benefits outweigh these risks. But the importance of a stable microbiome in early life should be a part of these considerations.

The rates of caesarean section are increasing in many high-income regions and this is associated with an increased risk of asthma. Similar findings were reached in several studies: 'meta-analyses' studies (combining data from all published studies)[78, 79]; a study of more than 1.8 million children in Norway[80]; a registered case control study in Finland;[81] and a US cohort.[82] This effect is probably because the operating theatre does not offer baby the normal 'innoculum' of healthy bacteria of a vaginal delivery. The infant gut microbiota resembles that of the mother's vagina, with progressive acquisition of the normal lactobacillus, bifidobacterium, and bacteroides species. Infants delivered by caesarean section initially acquire the microbial communities found on skin and the development of a more complex microbiota is slower. Swedish researchers have found that this is associated with reduced Type 1 immune responses during the first years of life.[83] This suggests a role for probiotics or other interventions (such as prebiotics) to counter the effects in high-risk deliveries. One study has shown that a probiotic cocktail (including both lactobacilli and bifidobacteria species) reduced the risk of allergies by almost 50 per cent in children delivered by caesarean section.[84]

Another possible risk factor for allergy (asthma) is conception by IVF. Neonatal complications and lower parental fertility appear to be factors,[85] possibly implicating environmental elements that affect reproductive health.

On the other hand, maternal consumption of foods (including peanuts) during pregnancy or lactation has now been found to *not* be risk factors for specific allergy in childhood.[86-89] More likely, environmental changes are altering early immune maturation to predispose to allergic responses. In other words, allergens should be viewed as the target of the allergic response rather than the cause.

INFANCY IS ANOTHER VERY BUSY TIME FOR THE IMMUNE SYSTEM: After birth, the immune system has to very quickly learn to distinguish 'friend' from 'foe' on a large scale. It does this by reading the proteins' antigens contained within each substance, and by looking for 'danger' signals. The vast majority of 'foreign' proteins need to be tolerated in the interests of our health and survival. Foods are the most obvious of these. Then there is the friendly gut 'symbiotic' bacteria vital for our survival and immune and metabolic health; we must actively promote its survival in our intestines so that more harmful species don't take over. Our friendly symbionts also provide the early critical cues for the infant immune system to mature. They take up residence in large numbers soon after birth in our intestines, where we also find the largest immune networks acting as an interface between our bodies and the external environment.

The human intestines harbour between 10 trillion and 100 trillion resident microbiota, and this vast and complex ecosystem forms gradually over the first years of life. The collective genetic material of these bacteria (called the 'microbiome') is estimated to contain 150 times more genes than our human genome. We are literally more bacteria than we are human. Through co-evolution and established 'mutualism' the microbiota play an essential role in the homeostasis of interconnected metabolic and immune functions. In the postnatal period, this is vitally important for normal immune maturation, with long-term implications for our immune health and a range of other NCDs.

Germ-free animal models reveal that without normal colonisation, there is failure of normal maturation and, in particular, of the systemic immune regulatory networks, resulting in both allergic and autoimmune phenomena.[90]

Differences in colonisation patterns in humans living in high- versus low-income countries result in variations in early immune diseases, such as allergy and autoimmune (Type 1) diabetes (Chapter 5). This again underscores the likely importance of strategies that improve gut homeostasis and the microbiome as part of broader disease prevention strategies.[91]

Postnatal colonisation is also affected by breastfeeding, infant dietary patterns, antibiotics and a range of other environmental factors. One promising postnatal intervention has been the use of soluble fibre (known as prebiotics or oligosaccharides) that selectively stimulates the growth of beneficial gut microbiota, particularly bifidobacteria but also lactobacilli.[92-94] These bacteria ferment oligosaccharides to form short-chain fatty acids (SCFA) that have direct anti-inflammatory effects.[95] This is very important for gut health, promoting the intestinal barrier function, and reduces systemic leakage of bacteria products such as endotoxin. SCFA stimulate the development of regulatory immune responses. Clinical trials of prebiotics in humans show beneficial effects on the microbiome and immune function, with reduced systemic inflammation, plasma lipids and markers of metabolic syndrome in overweight adults.[96] In young children, studies show beneficial effects on early colonisation and a reduction of eczema with pre-biotic supplementation.[97, 98] There are several studies now underway to explore this in more detail.

Probiotics have been far more extensively investigated in infancy and late pregnancy for allergy prevention. Collectively, these studies suggest a protective effect on eczema[99] and a combi-nation of prenatal and postnatal probiotic supplementation appears most effective. However, there has been no consistent protection from other allergic outcomes. So far, research has mostly focused on lactobacilli and bifidobacteria strains, and future studies are anticipated to determine if 'next-generation probiotics', such as stronger butyrate and proprionate producers and immunomodula-tory *Bacteroides* strains, have more powerful effects.

Helicobacter pylori: another old friend?

Since Barry Marshall and Robin Warren discovered H. pylori as the cause of gastric ulcers at our University in 1983,[100,101] we have also begun to understand that its presence in the stomach does not necessarily indicate pathology. In fact, H. pylori is turning out to be another old friend. We have a history of co-evolution with H. pylori dating back some 60,000 years. In traditional environ-ments we are colonised with H. pylori from around four to six months of age, and it persists in the stomach for the rest our lives unless we eradicate it with antibiotic treatment. It now appears to be important for normal immune development. Indeed, early colonisation with H. pylori appears to reduce the risk of allergic disease.[102-104] And disease appears more likely if colonisation is delayed or does not occur at all.

A level of interdependence with H. pylori may have evolved, and it is emerging as an important initiator of immune regulatory function in the upper gastrointestinal tract.[105] By skewing our adaptive immune response towards immune tolerance, it evades our innate and adaptive immune responses. At the same time, the induction of our regulatory responses inhibits autoimmune and allergic T-cell clones, conferring benefits through protection against allergies, asthma, and inflammatory bowel diseases.[106] H. pylori are also continuously shed in large numbers from the stomach into the lower intestine where they may also modulate immune function.

In animal studies, H. pylori 'infection' efficiently protects mice from airway inflammation, allergic responses and asthma, particularly when colonised in the newborn period. This effect appears to be due to induction of regulatory T cells (T-reg) and was lost when H. pylori was eradicated by antibiotics.[107]

We may learn to harness the properties of H. pylori for the prevention and treatment of not only allergic and autoimmune diseases, but also potentially other NCDs. We are now teaming up with Professor Barry Marshall to explore a possible role for attenuated strains of H. pylori for treating and preventing disease in infancy. This will be a long road with many safety hurdles, aiming to return to somewhere close to our traditional relationship.

INFANT FEEDING: IS THERE A 'WINDOW OF TOLERANCE' FOR REDUCING ALLERGY?

Infant feeding has become a major question in allergy – and a great source of confusion for both the general public and healthcare professionals. The allergy field is facing the mounting evidence that delayed complementary feeding and allergen avoidance may actually increase allergy risk.

Complementary feeding before three to four months is associated with an increased risk of allergic disease, presumably because of immaturity of the gut and related immune system, and lack of established microbiome.[108] But the risk also increases if complementary feeding is delayed until after the age of six months. This has lead to the speculative window of somewhere between four and six months of age to introduce complementary foods. These concepts are based on only observational data, meaning that the level of evidence is not high.

None of this negates recommendations to continue breastfeeding for as long as possible. Allergens are secreted normally in mother's milk, and appear to an important early source of exposure. This may actually be important in initiating, maintaining and reinforcing normal tolerance to foods and even inhaled allergens. When breastfeeding is not possible, the use of hydrolysed formulas appears to confer some protective effect compared with normal cow's milk–based formulas, though more studies are needed, and the protective effects are not large.

Regular exposure to ubiquitous proteins may promote subsequent clinical tolerance. This may be why allergen-avoidance strategies (delayed exposure) may have actually contributed to an increased allergic risk. The most recent cohort studies show the delayed introduction of specific foods (oats, wheat, cow's milk, fish and egg) is associated with increased allergic sensitisation and allergic disease. This gives continued support for the current recommendations by expert bodies[109-111] that there is *no evidence* that delayed introduction of complementary foods *beyond* four months of age is beneficial.

Many countries currently have two sets of recommendations around the timing of introducing 'solid' foods: on one hand parents concerned about allergy prevention are advised not to delay complementary foods (i.e. to start complementary foods between four to six months of age), and on the other hand the WHO recommends exclusively breastfeeding with delayed complementary feeding foods until after six months of age; see Chapter 4. This is clearly unhelpful to parents. The problem remains that we need better evidence before definitive statements can be made in either direction. This is likely to take a few more years yet, as the results of the clinical trials come to hand.

Similarly, attitudes and practices over maternal nutrition and the consumption of allergenic foods remain in disarray. While randomised controlled clinical trials, including our own, will address the role of allergen exposure in the early postnatal period, there appear to be no clinical trials to address this issue in pregnancy.

Studies show that a significant proportion of infants *already* have food sensitisation and clinical reactivity (including anaphylaxis) *prior* to the 'first' introduction of foods at four to six months[112] indicating that earlier preventive interventions will ultimately be required. And this may not necessarily mean using allergens at all, but other immunomodulatory strategies. Again, if allergens are the 'target' and not the 'cause' of the allergy epidemic they may not hold the final solution.

THE WIDER IMPLICATIONS OF EARLY 'IMMUNE HEALTH':

In almost every chapter of this book we have seen that the immune system plays an integral role in many aspects of health, as well as many of the pathological processes that lead to disease.

From the earliest moments of life, the immune system allows the rich and complex interactions between mother and foetus. It governs the complex processes involved in brain development and the refinement of our neural networks (Chapter 7). It allows us read, respond and adapt to our circumstances, predicting and anticipating the level of threat while carefully distinguishing and tolerating harmless things. And it even allows us to harvest energy supplies more efficiently, through complex relationships with the trillions of symbiotic bacteria that make their home in our intestines (Chapter 4).

The developing immune system is *designed* to respond quickly to our circumstances. That is how we survive. This may also be the reason that environment change has been associated with immune diseases so early in life. The 'allergy epidemic' is the clearest evidence we have that the early immune system is exquisitely vulnerable to modern lifestyle change. This can also offer us vital clues about why we are also increasingly prone to other NCDs, which all generally feature inflammation and immune dysregulation.

There is growing awareness that inflammation is an important part of the *initiating* events that lead to a range of disorders including cardiovascular disease, obesity and neuropsychiatric disorders, as discussed in detail in previous chapters. It is logical to wonder if the same environmental risk factors might be implicated in the rise in so many inflammatory diseases. Because eczema, food allergy and asthma appear so much earlier in life it is more obvious that early events *must* be important. Given that diet, microbes, exercise, stress and modern pollutants all affect the developing immune system, it can be no surprise that these are also linked with the rise in so many other NCDs. These are among the many lifestyle factors that are associated with the significantly higher baseline levels of inflammation that underpin many NCDs, and are the logical targets in preventing disease (Chapter 6).

And in closing this chapter I will borrow from my previous thoughts: that although immune disease is one of the prices we pay for modern living, we have every reason to hope that there are ways of overcoming this. The very fact that environmental change has precipitated such an increase in disease reveals the great plasticity of the immune system, as with other body systems. We can now hope to *harness that same plasticity* and use the environment to prevent disease before it develops, or to reprogram errant responses once they have developed. The answers are very likely to be right in front of us; we just have to understand them.

11

Healthy ageing starts in early life: genetics, epigenetics and telomeres

And so, from hour to hour, we ripe and ripe
And then from hour to hour, we rot and rot:
And thereby hangs a tale. [1]

WILLIAM SHAKESPEARE

We live by the same rules of biology as all living things. We are all slowly ageing. And as we get older, it becomes more obvious that some people age faster than others. The more we learn about the processes of ageing, the more we understand that our individual 'rate of ageing' is determined by factors that operate much earlier in life. Cellular programs that govern both our *development* and then our *decline* are still not well understood. A key goal of the 'DOHaD' field is to understand how these processes are related, and how we can improve both longevity, and health across the lifespan.

Absurdly, our society has become focused on the appearance of health when all logic dictates that we might achieve this anyway if we actually become healthy on the inside. The real goal and universal hope is that we might find a way to genuinely improve the human 'health-span' – both our longevity and the quality of our health as we age. As we learn more about the biology of ageing, this no longer seems impossible. It must be remembered that our greatest challenge may lie not in our biology, but in overcoming the human behaviour that is driving the widening gap in life expectancy between the wealthy and the poor. To make a truly meaningful change for humanity, improving 'health-span' and longevity must be something that is within reasonable reach of as many people as possible.

WHAT IS AGEING?

The biology of ageing needs to be considered at both the 'cellular' level and at the more general 'organism' level of whole-body functioning. Cells in different organs grow and divide at different rates. As cells naturally age and die, they are replaced. At the peak of our reproductive years (generally between twenty and thirty-five years) our regenerative capacity is nearly perfect, but after that it gradually deteriorates with a slow decline in our ability to respond to stress, rejuvenate and maintain our metabolic balance. As a result we also become more prone to disease and there appears to be a slow fall-off in our many organ functions, including our cognitive abilities. A decline in our immune function also contributes to the deteriorating function of many systems, as well as our susceptibility to cancer and degenerative disease.

As senescent cells (cells which no longer replicate) accumulate in various tissues and organs, they influence the structure and function of the tissues around them, and contribute to the age-related degeneration seen in various organs, even in foetal life. When senescent cells are removed from the tissue experimentally, the ageing process can be delayed, at least in mice.[2] Removal of cells by 'self-eating' or 'autophagy' is part of how tissues recycle, keep house and maintain normal functions. It has been proposed that if these functions become impaired, ageing and the risk of age-related diseases increases.[3] For example, a protein that inhibits autophagy (called mTOR) has been linked with ageing.[4] A number of lifestyle factors, such as high-calorie diets and reduced physical activity may exert some of their effect through mTOR. This may be why, at least in some animals, calorie restriction appears to increase longevity and health-span, and why some drugs that inhibit mTOR also increase lifespan,[5] as discussed further below.

A number of elements are considered important for successful ageing. Firstly, avoiding disease or disability. Secondly, high cognitive and physical function. And thirdly, active engagement with life and community.[6] It does not seem unreasonable or impossible to strive for these. Having such goals is the first step in achieving them.

So why doesn't the process of renewal and replacement continue forever? Is ageing an unavoidable consequence of carbon-based life? How much of our lifespan is coded in our genes and how much is determined by our experiences and our environment?

There have been a significant number of theories on this, and not all are mutually exclusive. External factors that have been proposed to affect ageing include radiation, smoking and environmental toxins, high caloric load, chronic infections, psychological stress, and isolation (below). There are a number of theories on how this may translate to cellular damage and senescence, including the accumulation of 'free radicals'[7] and other waste products that may damage DNA or interfere with cellular functions. The hormonal changes that occur with reproductive decline are also likely to contribute to changes in tissue and cellular functions with ageing. At a more general level, it has been proposed that 'wear and tear' simply accumulates over time. Although we have an inbuilt capacity to repair this damage, there may be a trade-off between being able to repair damage quickly in the interest of immediate survival, and the risk of minor errors during the repair. This potential accumulation of errors over time is basis of a 'misrepair' theory of ageing.

There is strong rationale for the theory that DNA damage and errors in DNA repair are central to the ageing process.[8] Increasing age is one of the strongest risk factors for cancer, which becomes almost inevitable with very advanced age. Most of the known causes of cancer also induce DNA damage (Chapter 9). And this includes the list of factors that have been implicated in accelerated ageing. Paradoxically, the malignant transformation that occurs in cancer cells leads to immortality for those cells, with the loss of the in-built ageing, self-regulation and programmed cell death. This aggressive and destructive form of cell survival is very different from the longevity we strive for, but it has revealed genes that are important for cell death and cell survival. Using this knowledge in more short-lived species such as mice, worms and fruit flies, scientists have been able to manipulate these genes experimentally to prolong life. While this cannot be extrapolated to humans, it has shown that genetic alterations that increase DNA repair, and reduce oxidative stress and 'programmed' cell death (apoptosis), can extend lifespan quite substantially in some species.

A BIOLOGICAL CLOCK:

Even before we understood something of the biology of ageing, we could surmise there must be some 'blueprint' to determine lifespan, as longevity is clearly very different for a tree or for a human than for a worm. Much of this appears to be determined by our genetic code. Our genes also determine the many individual characteristics that are so obviously inherited from our parents. But we are fast learning that our environment and our life experiences also play a substantive role in longevity and disease disposition.

It is still believed that, like most living species, we have an 'upper limit' on the number of times our cells can divide. In 1961 Leonard Hayflick showed that normal human cells do have limited capacity to divide,[9] challenging an emerging dogma that cells might divide 'forever' given the right conditions.

Hayflick's experiments showed human embryonic cells only divided around forty to sixty times before they entered a state of cellular ageing or 'senescence'. Beyond that point if the cells survived, they changed shape, activity, and function and no longer divided. This later became known as the 'Hayflick Limit'[10] and this is thought to govern the maximum lifespan of each species. Hayflick and his team calculated that only fifty-four divisions would be required from a single cell to produce the total output of human immune cells (leukocytes) in a sixty-year period.[11]

It would also mean that the remaining capacity for cell division would depend on the age of the individual from which the cells were taken. This proved to be the case. There were also anticipated differences between species that appear to correlate with their maximum lifespan. Cells from the very long-lived Galapagos tortoise have the capacity to divide ninety to 125 times before they go into senescence, whereas cells from a mouse will only sustain fourteen to twenty-eight divisions.

Even so, these experiments focused on specific cell types rather than whole organisms, and it is clear that cells in different tissues divide at different rates. Hayflick proposed that the *root causes of ageing occur as a consequence of decrements in some few cell types in which the rate is fastest and the effects greatest.*[12] In other words, we only live as long as our most vulnerable tissues.

It was clear that cancer cells could 'escape' this program of decline, pointing to some form of genetic change. Hayflick proposed that *normal* cells contained some form of cellular counting mechanism that he called the 'replicometer'. Even when cells are cryogenically preserved, they seemed to remember this when they were woken up again.

In 1978, Australian researcher Elizabeth Blackburn discovered that this illusive cellular countdown mechanism resides in the telomere cap, which sits at the end of each of our chromosomes.[13] This discovery, and her subsequent contribution to understanding telomeres and ageing, earned her a Nobel Prize in 2009, which she shared with her graduate student, Carol Greider.

Do telomeres hold the keys to ageing and our biological clock?

Telomeres are long, repeated segments of DNA at the end of chromosomes. As we age, our telomeres become shorter – every time cells divide. Telomere shortening has been associated with a many age-related diseases and their risk factors,[14] although this shortening is a less-accurate measure of ageing than our actual chronological age.[15] Nonetheless, there is still great interest in whether our telomeres might one day offer a predictive measure of life expectancy, and a possible strategy to extend it.

If we start life with shorter telomeres, we have less 'room to move'. At any given age, our telomere length will first depend on the initial length of our telomeres at birth (below) and the degree of telomere erosion that occurs thereafter.[16] The rate of 'erosion' will in turn depend on the cumulative exposure to events that produce inflammation, oxidative stress (free radicals) and DNA damage, as well as the activity of the telomerase enzyme to counter this. There are a number of adverse events in early life that will prematurely shorten our telomeres, and most of these are also associated with NCDs and reduced life expectancy, as we will see a little later.[17]

So why do our telomeres shorten? Each time a cell divides, it must replicate its DNA and give its daughter cells exact copies. During this process, the double strands of DNA are separated and an enzyme called 'DNA polymerase' runs along the length of each strand to generate a mirror sequence using 'nucleotide' building blocks. When it gets to the end of the strand, there needs to be something to stop it unravelling and to stop it sticking onto neighbouring DNA strands from other chromosomes. That is what a telomere does: like the 'bottom stop' on a zipper. But every time a cell divides the telomeres get shorter. This is because when the DNA polymerase enzyme gets to the end of the chromosome it can't replicate the small length of DNA (part of the telomere) that it is sitting on. Without telomeres, our genes would progressively lose important information as DNA is truncated during each cell division. Our telomere buffer appears to be between 10,000 and 15,000 base-pairs (DNA nucleotides) in length, although this varies.[18] It has been estimated that we lose between thirty to 200 base-pairs with each cell division. So we have a fair bit of room to move!

CAN WE EXTEND TELOMERES?

In 1973 Russian theoretical biologist Alexey Olovnikov first proposed that there must be a compensatory mechanism for telomere shortening.[19] Then in 1985 Elizabeth Blackburn and her graduate student Carol Greider published evidence supporting this theory, describing the telomerase enzyme in *Tetrahymena* protozoa, and revealing how this synthesises and elongates telomeres.[20] Not long after this, telomerase was found to be much more active in cancer cells and in immortal human cell lines.

Another breakthrough came in 1998 when it was shown that artificially inducing telomerase activity in normal human cells in the 'test tube' allowed the cells to maintain their telomeres' length and extend their normal lifespan by more than five times.[21] This demonstrated a clear role of telomerase in delaying cell senescence. Scientists have now engineered mice that are completely deficient in telomerase, so they can then restore levels to see what happens. Doing this, researchers at Harvard Medical School have essentially shown that they can reverse the ageing process.

Mice that lack the telomerase enzyme age prematurely. They are barely fertile and suffer from age-related conditions such as osteoporosis, diabetes and neurodegeneration. And they also die young. Restoring telomerase induced a dramatic reversal of the ageing effect: shrivelled, degenerated organs rejuvenated and gut, immune, and metabolic functions recuperated.[22] This has obviously generated excitement about both reversibility and the potential for boosting telomerase in humans to delay ageing.

But most scientists also urge caution. These experiments reveal more about the function of telomerase, but we cannot make too many assumptions about how this might apply to humans. There have been concerns that strategies to increase telomerase might increase cancer risk. This has prompted research into telomerase *inhibition* as a cancer therapy. Others disagree with this and argue that telomerase should *prevent* healthy cells from becoming cancerous in the first place by preventing DNA damage.[23] Cancer cells certainly don't play by the rules, which is why there is understandably both caution and disagreement.

Interestingly, in 2012 researchers gave mice a telomerase boost and saw dramatic benefits without any increase in cancer. They reported 'remarkable beneficial effects on health and fitness' with improved metabolic health, bone density, and coordination. Lifespan was also significantly increased. The effect on lifespan was greater (a 24 per cent increase) for the animals treated at an earlier age (one year old) compared with those that were treated in old age (two years).[24] But even the older ones still showed a 13 per cent increase in lifespan.

It is unlikely that the secret to longevity would lie in just one element of our biology. Ageing is clearly a highly complex process, involving many and varied factors. And, therefore, it also seems unlikely that this might be the elusive secret elixir, or the fountain of youth.

Putting all this together, it so far appears that telomeres generally get shorter as cells age because telomerase levels become insufficient to compensate for the normal loss of telomeric DNA with each cell division. This process may be accelerated during periods of stress, and may also be compounded if telomeres are already shorter as a result of adverse early life events.

LINKS BETWEEN TELOMERE LENGTH, LONGEVITY AND DISEASE IN HUMANS?

Dyskeratosis congenita (DKC), a very rare genetic disorder, reveals what happens when telomerase is not functioning normally in humans. This condition is caused by a mutation in the gene that codes for part of the telomerase enzyme. It results in extremely short telomeres and signs of premature ageing, often from childhood, with premature greying, skin damage, immune failure and a greatly increased risk of cancer. Those with more severe forms of the disease have significantly reduced life expectancy, confirming the importance of telomerase in humans.

A study on how telomeres shorten with age was published in *Nature* in 1990, focussing on cancer.[25] The researchers also found that the average telomere length was shorter in older people, and that telomeres were much longer in foetal tissues and in sperm than in adult blood and body tissue. Based on this they proposed that telomerase might be inactive in many tissues, and that telomere length reflects the number of cell divisions that have already taken place in forming that particular tissue. In other words, as normal cells age, telomeres get shorter because telomerase levels are insufficient to compensate for the normal loss. During periods of biological adversity, a further unfavourable shift in this balance may accelerate this loss – at least that is the theory.

Almost twenty-five years later, there have been a large number of studies in many human populations. One of the most common, non-invasive ways of measuring telomeres has been in immune cells in circulating blood. There are limitations here. While circulating immune cells may reflect the state of other cell types, telomere length varies between tissues and these differences are not measured using this method. And at any moment our telomere length reflects a complex combination of many inherited and environmental differences. How *changes* in telomere length vary over time in the *same* individuals is likely to give us the best information, and to examine this we need a longitudinal study. Studies of serial sampling are now underway, but obviously take time. And complex genetic and environment influences will remain difficult to separate, although new statistical modelling may be providing some possible solutions.

So, with these caveats and limitations in mind, what has been found so far?

First, there now seems little doubt that our telomere length decreases with age. It actually appears that telomere shortening is extremely rapid *early* in life, and then slows down considerably during adulthood. In a recent systematic review of 124 cross-sectional studies and five longitudinal studies, the data from the five longitudinal studies, which performed *serial* sampling, showed a higher average annual loss of between thirty-two and forty-six base pairs (the nucleotide building blocks of DNA), compared to twenty-five base pairs in the other studies. Clearly, this also reveals a level of variability between individuals, and between populations. There have been other anomalies and the subject needs to be further studied and understood.

There are some factors that seem likely to determine some of the *variability* between individuals, but these are still not well studied. For example, Gardner and her team reviewed thirty-six cohorts with a total of 36,230 participants and confirmed that females have longer telomeres than males.[26] The fact that there are no sex differences in telomere length at birth suggests that the differences between males and females emerge as we age.[27]

Then, we come to the question of how telomere length predicts or associates with *specific* age-related *diseases*. Accelerated telomere erosion has been associated with many metabolic and inflammatory diseases, although the nature of this relationship is far from clear. In almost all studies, shorter telomeres have been associated with increased risk of disease or adverse lifestyle factors (as summarised recently by Bojesen).[28] One of the most consistent relationships has been in cardiovascular disease, with associations between a shorter mean telomere length and the disease. [29] Shorter telomere length is also associated with asthma, chronic lung disease and declining lung function.[30, 31] Associations for diabetes, metabolic syndrome and obesity are less strong.[32, 33] Telomere attrition has also been associated with cognitive decline in the elderly.[34] Several systematic reviews, of more than 25,000 people, found associations between short telomeres and cancer.[35, 36] However, a Danish study following 47,102 of the general population for twenty years found that although short telomere length was associated with reduced survival after cancer, it did not predict cancer risk.[37]

Finally, we come to the question of whether telomeres have any *practical* role in the clinic for predicting longevity, or as an accurate biological measure of ageing. Here the answer is fairly clear – they don't have a role, at least for the foreseeable future. Telomere length falls a long way short of reaching the required criteria for a well-defined biomarker of ageing developed by the American Federation of Aging Research.[38] In particular, there are no defined 'normal' levels. And we can't be sure whether it is measuring the underlying ageing process and not the effect of disease. Bojesen likened telomere length to 'a soup cube' carrying condensed information about past exposures and future health.[39] Even if this does one day prove to be a crystal ball, we are far from being able to read it.

That said, this remains a fascinating area of research. But the role of telomeres and of telomerases in ageing is far more nuanced than first thought. Elizabeth Blackburn summarised this in an interview:

>...*telomere changes are associated with everything from cardiovascular disease ... diabetes risks such as insulin resistance, vascular dementia, to osteoarthritis. The list goes on and on and the correlation is always in the same direction: shorter telomere length is associated with more*

disease. The association is absolutely solid now because it has been found in so many cohorts that it cannot be a statistical accident. What does this mean? That's a really exciting question and I don't know yet. I am sceptical that the answer is going to be very simple.' [40]

LIFESTYLE RISKS, TELOMERES AND AGEING:

Telomeres are only part of the story of ageing. So many of the factors implicated in ageing are also implicated in the many specific NCDs discussed in this book. Some of these have also been specifically associated with telomere shortening, including smoking, obesity and dietary patterns, physical inactivity, poverty, chronic low-grade inflammation, sleep patterns, and even marital status. But the findings are not all consistent, and these adverse exposures have many other biological effects that may accelerate ageing. Apparent effects on our telomeres may be merely a marker of their general impact, rather than being the direct cause of the ageing process.

That said, there is a strong biological basis to suspect that the obvious oxidative stress induced by factors such cigarette smoking might have some direct effect on telomeres. In one study, smokers had almost 20 per cent greater loss in telomere length than non-smokers.[41] And the more they smoked, the greater the loss. Other studies found similar relationships.

Obesity is another important risk factor for age-related diseases. In one study, telomeres of obese women were on average 240 base-pairs shorter than telomeres of lean women.[42] This could fit with the many metabolic changes that might accelerate oxidative stress and the ageing process. Subcutaneous fat, body fat and the obesity-associated hormone, leptin (Chapter 4), have all been associated with shorter telomere length.[43] There has been great interest in the role of caloric burden in ageing and, conversely, how caloric restriction can increase longevity, at least in animal studies (as discussed further below).

There is also emerging evidence that physical activity may protect against telomere shortening. This was seen in a study of 2,401 twins.[44] And poor sleep is another factor implicated in ageing. Shorter sleep has been associated with shorter telomeres in older men.[45] A similar association has been seen for women under fifty years but not in older women.[46] Sleep quality and rotating night shifts have also been linked to shorter telomeres in women.[47]

Inflammation, a recurring theme in NCDs and ageing, has also been examined in relation to telomere length. Levels of inflammatory cytokines such as interleukin-6 (Chapter 10) and measures of oxidative stress in the blood have both been associated with shorter telomeres in humans.[48, 49] These relationships have been seen in healthy volunteers as well as people suffering from NCDs such as cardiovascular disease and depression.[50] In a study of almost 2,000 US adults (aged seventy to seventy-nine years), telomere lengths were significantly shorter in those with elevated interleukin-6 or other inflammatory cytokines – although there was no relationship with C-reactive protein (CRP).[51] Inflammation and oxidative stress in foetal life is also associated with shorter telomere length, which may contribute to this risk of premature senescence (below).[52]

Psychological stress, anxiety and depression appear to be additional, separate risk factors for telomere shortening, beginning very in early life.[53, 54] Activation of stress hormonal responses (Chapter 7) is associated with immune dysregulation, inflammation and many NCDs. This has been linked with accelerated cellular and organismal ageing, higher levels of oxidative stress (free radicals), shorter telomere length, and reduced telomerase activity in immune cells.[55] Depression, bipolar disorder, perceived stress and experiencing major loss have all been associated with greater telomere shortening. In one study, the telomere length of people who had experienced depressive symptoms for more than ten years cumulatively was 281 base pairs shorter than that in those without depression.[56] This corresponds to approximately seven years of 'accelerated' cell ageing. Because telomere shortening does not antedate depression, it is more likely to be a consequence than a cause of the disease. In men, 'hostility' has also been associated with a significantly increased risk of age-related disease and mortality, and recently British researchers showed that men with more hostile behaviour have shorter telomeres and increased telomerase activity.[57]

Low socio-economic status is a well-known risk factor for shortened life expectancy, and this has also been associated with shorter telomeres, over and above the effects of associated risks such as smoking and obesity, though some studies have failed to substantiate this. In older adults, being married, as well as having a higher income, have been associated with longer telomeres.

Other associations include low vitamin D levels, suggesting another potential mechanism of the disease-protective effects of vitamin D, and long-term hormone replacement therapy in menopausal women, which appears to protect from telomere shortening.[58]

The bottom line is that most of the factors linked with age-related diseases have been shown, at least in some studies, to have some relationship with telomere length, and most have been in the expected direction. These are already known targets for reducing disease, whether or not the relationship with telomeres is eventually proven.

HEALTHY AGEING – HOW MUCH IS IN OUR GENES?

Many people are surprised to learn how few genes that humans carry. Only a few more than a chicken or a fruit fly. Early estimates of 6 million genes were revised down to 100,000 by 1990, and as we learned more about the human genome estimates fell to somewhere around 20,000 genes. There were also unexpected differences in the numbers of genes between people.[59, 60] We all have subtly different versions of many genes, and these play an important role in how 'well' we age.

So far, the longest surviving human on record, a French woman named Jeanne Calment, lived for 122 years. It is unlikely, but not impossible, that our genetic potential is much more than this. The fundamental question is: how much of our genetic predisposition can we hope to change?

There are likely to be more than 400 genes that interact to maintain telomere length. Our telomere length is strongly determined by our genes, and experts have estimated that somewhere between 35 and 80 per cent of the variation in telomere length is inherited.[61, 62] And then there are thousands of other genes that govern the many other aspects of our growth, development and senescence. Most of the heritable variability in the general population is caused by subtle genetic variants (genetic polymorphisms) called 'alleles'. This is true of virtually all common conditions, including the many NCDs we have considered so far.

So what are genetic polymorphisms? Each gene is a sequence of DNA nucleotide building blocks that hold that code for a particular protein. Some changes in sequence can cause the protein to be malformed or not made at all, and these detrimental versions tend to be gradually reduced in populations during the natural-selection process of evolution. But the very minor differences in gene sequence, which do not greatly affect function, will persist in a population. One of the best-known examples of this are the different 'blood groups'. Other alleles influence the differences in our height and skin colour, without any significant effect on our disease risk.

Alleles may be as minor as only one substituted nucleotide in the DNA sequence – called 'single nucleotide polymorphisms' (SNPs; 'snips'). Not all 'SNPs' produce a functional difference in the protein they produce, but some slightly change the risk of disease under certain conditions. These alleles influence our susceptibility to heart disease, obesity, diabetes, dementia, allergy, and virtually every common disease. They also influence our behaviour, appetite and propensity for anxiety and depression. Most importantly, this predisposition is not absolute and is usually dependent on the environment. Obesity predisposition, for example, manifests only when food is overabundant. The surge in 'modern' NCDs is the result of these kinds of gene–environment interactions unmasking dormant tendencies for inflammation and disease.

The individual differences in disease patterns will also depend on other susceptibility genes: some people will be predisposed to heart disease and others to diabetes or Alzheimer's disease. We can already see that a significant proportion of susceptible genes identified in many NCDs lie in immune pathways – and differences in the propensity for inflammation is an important factor in disease predisposition.

So why have more 'inflammatory' alleles been conserved through evolution if they are causing so much disease today? In particular, why are certain populations in the developing world much more prone to disease in the Western environment?

For over twenty years, in the office next to mine, my colleague Professor Peter Le Souef has been investigating genetic polymorphisms that might predispose to asthma. He and his team have been travelling the world, comparing alleles in inflammatory genes in people from different regions. We might have expected that Caucasians living in Australia, suffering *more* asthma, might have had *more* 'pro-inflammatory' alleles than people with no asthma living in the tropics. But they found the complete opposite.[63] The pro-inflammatory alleles were actually more common in populations with long-term tropical ancestry than those with long-term residence in temperate regions. This was seen in several genes involved in inflammation. In 2000, they published the following theory in *The Lancet*.

Because modern man's ancestors lived in a hostile environment where infectious, tropical diseases were rife, this would have favoured genetic selection for pro-inflammatory immune responses to aid in their defence. But, on migrating to temperate regions, vigorous inflammatory responses would have been less important and selected against due to the potential unnecessary harm they may cause. This could explain why people in our modern environment who have tropical ancestry have a higher incidence of inflammatory diseases than those with temperate ancestry. Although Le Souef's focus was on asthma, this applies equally well to what we see for virtually every inflammatory NCD. It also shows how evolutionary adaptation in the human immune system can explain many of the differences in disease susceptibility that we see today.

ARE THERE GENETIC VARIANTS THAT INFLUENCE TELOMERES AND AGEING?

We know that telomere length is inherited. But knowledge of the common genetic alleles, and their role in the variable regulation of telomere length in the general population, is still evolving. And 'gene networks', likely as important, have not been sufficiently examined.

There are several approaches to identifying genes that may be associated with a biological process. One traditional method is to look at specific 'candidate genes' in a known pathway – immune genes in allergy, or telomerase genes in centenarians, for example. Researchers then compare whether one allele of the candidate gene is more common in people with the condition of interest than those without it.

Using this approach, researchers found that centenarians and their offspring had higher frequencies of certain alleles associated with telomerase enzyme activity (i.e. particular versions of these genes, including '*TERT*' and '*TERC*') than offspring of parents with survival of around seventy years.[64] Even so, these and other common variants explain less than 1 per cent of the variance in telomere length.[65] This 'candidate' approach is useful, but it narrows the field and does not allow for the networks of genes that might also have an effect.

The other approach comes from 'genome-wide association studies' (GWAS) which examine the *entire* genetic sequence and look for patterns of difference in large numbers of people, again comparing people with a condition to those without. The GWAS approach looks at *all* possible SNPs across the entire genome, searching for any that are more or less likely to be present in the people with the condition of interest. This obviously throws up thousands of data points, but the ones with strong relationships stand out very easily. Above a certain level, SNPs are significantly associated with a disease, and the higher they are, the stronger the association. But these are only associations, and may not be the direct cause of the disease.

The critical thing is that there is an accurate measure of what we are comparing. And that is the biggest challenge when we look at conditions such as obesity, depression, dementia, asthma and heart disease, as these are heterogeneous – variable between the people that have them. Studying large numbers, preferably thousands, of people with these conditions helps iron out some of this uncertainty.

Using the GWAS approach, researchers have identified a number of genes that are highly associated with telomere length, confirming '*TERT*' and '*TERC*' among them. Importantly, these SNPs were also associated with age-related diseases such as several cancers, lung disease and heart disease, lending more support to the idea that there might be a 'causal role' between telomere-length variation and these diseases.[66]

But again, telomeres are only part of the ageing and longevity story. Other GWAS studies have attempted to examine measures of longevity in general, and this has been more challenging.[67] So far only one gene, the *APOE* gene, has been consistently found to be associated with longevity in studies of centenarians.[68] This gene codes for a protein called 'apolipoprotein E' which removes excess cholesterol from the blood. *APOE* has also been associated with specific age-related diseases such as Alzheimer's disease and cardiovascular disease. Some studies have shown that polymorphisms in the insulin growth factor-1 (IGF-1) and its receptor may be linked to longevity.[69] So, although it is estimated that about 25 per cent of the factors that determine our age at death are inherited, it has been disappointing that not more genes have been identified. As yet, there is no strong genetic measure that accurately links to telomere length or, more importantly, longevity. And it is unrealistic to expect that there will be, given the vast numbers of processes and gene networks that contribute to the ageing process.

EPIGENETIC CONTROL – OUR 'NON-CODING' DNA IS NOT ALL JUNK:

'Epigenetics' basically refers to the normal mechanisms that control whether each particular gene is switched on or off and, as touched on earlier, the 'epigenetic control' of gene expression might be just as important as the actual DNA code itself.

Our DNA provides the 'blueprint' template for *everything* that we are able to make and produce throughout every stage of life. It therefore also codes for the 'epigenetic' machinery it needs to control its own gene expression. These control sequences are hidden in the vast 'non-coding' regions of our DNA that have been labelled as 'junk' because they don't code for proteins or other known functional elements. But there are thousands of regulatory genes within these regions that don't code for proteins (non-coding RNAs) that can regulate cell functions through epigenetic effects – by altering chromatin structure, or by controlling the production of proteins once their genes are switched on.[70] This all sounds very complicated, because it is. We are still on the threshold of understanding the secrets of our DNA.

The environment can change our genes and our biology – not by changing the gene sequence, but by changing which genes are expressed. So even if *identical* twins have *identical* DNA with the *identical* alleles, they may have subtly *different* patterns of gene expression and disease if they experience different environments.

Our 'epigenetic program' coordinates *what*, *when* and *where* genes are expressed. Without this we would be a disorganised mass of cells that could not survive. Epigenetics also provides a clear explanation of how environmental changes can affect gene expression and alter early development in ways that can lead to future predisposition to many types of diseases (Chapter 4).

EPIGENETICS, DEVELOPMENT AND AGEING:

Our epigenetic program is fundamental to all aspects of our development, driving the differentiation of so many different tissues and organs, all from a single fertilised egg. This is also quickly emerging as a important dimension of ageing and longevity.

While the epigenetic program is inherited, it can be *easily influenced* by environmental exposures. While this may be adaptive, and 'designed' to allow the developing organism to alter its development in anticipation of its future environment, some exposures can also work against us and lead to increased risk of disease (below, and see, mismatch, in Chapter 4).[71] Embryonic and foetal life are particular times of vulnerability, when patterns of future gene expression may be set. At the other end of the spectrum, centenarians show delayed age-related methylation changes, and they can pass this ability for self-preservation on to their offspring (below).[72] While this is likely to reflect a lifetime of diverse influences, some of these patterns appear to be determined in early life.

In order to fit several metres of DNA, containing over 3 billion DNA base pairs, into the nucleus of each cell, the DNA is wrapped around 'histone' proteins – like a thread around a cotton-reel. These form bundles called 'nucleosomes' that look like beads on a string. All of these are then tightly packed on scaffolding proteins to form each chromosome, each with a telomere cap to protect the ends. This collective mass of genetic material, a mix of the DNA and scaffolding proteins, condenses to form chromosomes and is often referred to as 'chromatin'.

Chemical changes can 'open' the chromatin and DNA structure near a particular gene so that it can be expressed, or 'close' it so the gene stays silenced (these are 'epigenetic' changes). For example, the addition of specific biochemical structures to the DNA (such as 'methyl' groups) maintains the gene in a dormant state until it needs to be expressed. Then, when it needs to be activated (to make, for example, a hormone, a cytokine or a structural protein), the epigenetic program removes these 'methyl' groups so the gene can be expressed. The addition of 'acetyl' groups to histone tails is another common example of epigenetic change, which opens the chromatin to increase gene expression. Throughout life, these control mechanisms are finely orchestrated to maintain both the development and the function of our tissues. When epigenetic control mechanisms fail, genes escape this regulation, and we see cancers develop (Chapter 9). But more minor shifts in epigenetic control, induced by the environment, can also have implications for our disease predisposition.

We have already seen (Chapter 4), how exposing genetically *identical* animals to different environments in pregnancy can result in radically *different* appearances, eating behaviour and weight gain as adults. Some of these effects *only* occur when the environmental exposure occurs early in development, while tissues and biological responses are forming. But there is no doubt that postnatal exposures are also important. In twins, with identical DNA, gene methylation patterns can be quite different. These epigenetic differences increase with age and are greater in twins who have had different lifestyles.[73, 74] But even in twins raised together, these differences can begin to appear in early childhood, indicating the likely effects of subtle variations in the early environment.[75] This suggests that small-but-cumulative changes in DNA methylation occur in all of us over our lifetime, each adding to differences in disease susceptibility.

In our own laboratory, we have seen significant age-related epigenetic changes over the course of early childhood.[76] It has been suggested that as we age, there is a 'relaxation of epigenetic control' and variations in this epigenetic drift may influence our risk of ageing-related diseases.[77]

It is important to again emphasise that DNA methylation patterns vary between organs and tissues, and we are usually limited to examining blood samples. So while most of the human data are based on epigenetics in immune cells, this may not reflect what is happening in other tissues. This will remain a challenge.

The discovery of epigenetics has provided an exciting new insight into *how* recent environmental changes may have triggered such rapid changes in disease patterns, and may lead to strategies to modify gene expression to *prevent* disease. It seems likely that lifestyle practices that help us maintain a favourable 'epigenetic profile' will be important in longevity and healthy ageing, especially if they begin in early life.

WHAT CAN WE DO TO TURN BACK THE CLOCK?

The first question must be whether we *can* actually turn back the clock. It seems fairly unlikely. But we should try not to sabotage our potential too much! That is a more realistic goal.

As discussed in earlier chapters, healthy diet, regular physical activity, reducing stress, and maintaining a healthy weight and vitamin D levels all improve the quality of life and reduce disease, while smoking, high cholesterol levels, high blood sugars, high blood pressure, and greater waist circumference (visceral fat) are all things that we need to avoid. There is also every indication that promoting 'health' factors could have beneficial effects on our biological clock, in so far as most have been linked with longer telomeres or increased telomerase enzyme activity, at least in preliminary studies.

Taking a multi-faceted approach, one important study assessed the effects of comprehensive lifestyle change on telomeres and health. This included a diet of plant-based, low-fat, whole-foods; walking for a half an hour a day; using various stress-management techniques, including yoga, relaxation and meditation for an hour a day; and increased social support (one hour support-group sessions each week).[78] Volunteers exercised with an exercise physiologist, participated in stress-management sessions and a support group led by a clinical psychologist, and attended lectures from dieticians, nurses, and physicians. A control (comparison) group made only minor lifestyle changes. After three months the researchers found increased telomerase activity in the lifestyle-intervention group, measured in circulating immune-system cells.[79]

Examining the groups five years later, there was a signifi-cant increase in the telomere length in the lifestyle-intervention group, but a decrease in length in the control group.[80] The more the volunteers adhered to their lifestyle change, the longer their telomeres got. The participants in this study were all men, aged around sixty years, with low-risk prostatic cancer, and larger trials in less-selected populations are needed to confirm these findings.

A larger dietary-intervention study, aimed at weight reduction, saw changes in telomere length after five years on a 'Mediterranean diet'. People who had longer telomeres before they started the diet were more likely to lose weight, especially if their telomere length increased.[81]

The role of 'restricting' calories to increase longevity has been of great interest because of the striking effects in animals. In 1935, McCay and Maynard reported that restricting caloric intake by 40 per cent from the age of weaning dramatically extended lifespan.[82] There are now more than seventy years of data collected in various animal species showing similar, significant increases in lifespan. Rodents generally live 50 per cent longer with caloric restriction, with progressively less effect the later in life it is started.[83] Importantly, the goal of these diets has been to reduce calories while still *ensuring adequate intake of all essential nutrients, vitamins and minerals.* And beyond a certain point the effects are obviously deleterious on both health and lifespan.

Metabolic adaptations to short-term fluctuations in nutritional supply are normal, and part of many animals' biological strategies to cope with seasonal changes in food supply (Chapter 10). These may also be implicated in slowing down the ageing process. These changes include shifts in insulin and metabolic hormones such as leptin (Chapter 4), changes in immune function (Chapter 10), reduced metabolic rate, reduced oxidative stress and reduced body-temperature changes. More recently, caloric restriction was shown to increase telomere length in mice.[84] Caloric restriction also saw improved cardiovascular function, improved age-related decline in liver-tissue resilience, protection from age-related muscle loss, and trends toward increased collagen and elastic fibres in skin (although not in all studies).[85] In the brain, increased production of brain-derived neurotrophic factor (BDNF; Chapter 7), may increase the resistance of neurons in the brain to dysfunction and degeneration in animal models of neurodegenerative disorders.[86] In humans, one study has shown an increase in circulating levels of BDNF in obese adults after three months on 25-per-cent caloric restriction.[87]

Studies in primates, closer to humans, such as rhesus monkeys, have generated conflicting results. Some show similar beneficial metabolic effects to rodents, but not all studies have shown an increase in longevity.[88]

There are a number of ways that caloric restriction could influence ageing and longevity. A reduction in oxidative stress is one of the common explanations.[89] Caloric restriction also increases 'autophagy', which has been implicated in slowing down the ageing process (above). Autophagy is a normal process for recycling cell components and nutrients, and provides a critical source of nutrients during fasting. This capacity for autophagy declines with age, and is negligible by old age.[90] Because caloric restriction induces a state of 'activated autophagy' it appears to delay this age-related decline in animals. There is now interest in human drugs (such as rapamycin, which counters the effects of mTOR), that can mimic these effects by restoring or increasing the levels of autophagy.[91]

In one study, caloric restriction induced changes in the gut microbiota, increasing lactobacillus and other species known to have favourable metabolic and immune effects (Chapters 4 and 10).[92] This reveals the close connection between nutritional modulation of gut microbiota and healthy ageing, and could provide another previously unrecognised mechanism of the effects of caloric restriction or other dietary changes.

On the other hand, adverse effects of caloric restriction must be considered, particularly the negative impact on bone loss and the risk of osteoporosis, and reduced fertility. It is completely inappropriate in pregnancy, with adverse consequences for both mother and foetus.

Caloric restriction studies have also measured cardiometabolic risk, such as cholesterol, blood pressure, and insulin, with significant improvement in these risk factors.[93, 94] Controlled clinical trials have shown that long-term caloric restriction or exercise interventions improve cardiovascular risk.[95]

Several small-scale clinical trials have been published, including one with quite promising results. In 2006, Heilbronn and other colleagues in the USA performed a clinical trial of caloric restriction in overweight but non-obese adults.[96] There were four groups in the study: a control group; a group restricted to 75 per cent of their normal calories (i.e a 25 per cent calorie restriction); a group with a structured exercise program, but allowed to eat slightly more (12.5 per cent calorie restriction); and a group with very low-calorie diet of only 890 kcal per day until they achieved a 15 per cent weight reduction, then a weight maintenance diet. After six months, all of the intervention groups showed weight reduction (averaging a weight loss of 10 per cent in groups two and three), which was greatest in the very low-calorie diet (averaging 14 per cent weight loss). They also all showed favourable metabolic changes with lower insulin levels and oxidative stress as evidenced by DNA damage. Interestingly, the caloric restriction groups (including the exercise group) all showed a significant reduction in their energy requirements, over and above what could be explained or expected simply by their size reduction. This is consistent with the 'metabolic adaptations' seen in animal studies, supporting the idea that metabolic rate is reduced.

These beneficial effects could be achieved with relatively modest calorie restriction (between 12.5 per cent and 25 per cent depending on physical activity) and did not require a very low-calorie diet. It is vital, however, to examine the safety aspects of these studies. It is possible that even moderate calorie restriction may be harmful in lean people with low body fat, and it is certainly inappropriate in pregnancy and for children.

HEALTHY AGEING FROM CONCEPTION – IMPLICATIONS IN EARLY LIFE:

To really understand healthy ageing, we need to go back to the beginning of life. Paediatricians and geriatricians work at the opposite ends of the life-health spectrum, and we don't tend to work much together. If we did, we would get a much richer picture and a better chance of achieving healthy ageing. The 'DOHaD' approach to health offers a new forum to do this.

Telomere length in *early* life may offer a better prediction of potential lifespan than telomere length in later life, at least in animal studies.[97] We humans also show significant variability in the length of our telomeres at birth, and while some of this is inherited, it can also be influenced by adversity in foetal life – including maternal smoking, poor nutrition, infection and other causes of inflammation, and even poverty and psychological stress. It is proposed that accelerated cellular ageing during foetal life may foreshadow defective telomere maintenance, potentially increasing the risk of inflammation and NCDs in later life.[98]

California researchers found that children of mothers experiencing higher levels of psychological stress in pregnancy have shorter telomeres.[99] These effects were already evident at birth and were proportional to the levels of stress mothers experienced. Even by early adulthood they saw telomeres were shorter when mothers were exposed to severe stressors in pregnancy. This equated approximately to an extra three and a half years of cellular ageing. Adverse environments and social stress in later childhood have also been associated with shorter telomeres.[100]

We can't be sure that these events *cause* telomere shortening, or that telomere shortening *causes* the later differences in disease predisposition and life expectancy. In these cases, shorter telomeres may simply reflect the broader impact of an adverse environment in pregnancy on many aspects of cellular and metabolic programming. We also don't yet know how much of the telomere length at birth is determined at conception, and how much this may be altered over the course of the many cell divisions that occur during gestation. All of these things are likely to be important.

Studies have shown that parental factors before conception might have an effect. Older fathers tend to have children with longer telomeres, even though the telomere length in their other cells types is getting much shorter with age. There is even evidence that this might accumulate over generations, if paternal grandfathers were older when fathers were conceived.[101] This could just reflect a 'survival of the fittest' sperm producing stem cells, or there could be some other evolutionary purpose.[102] We just don't know yet. The implication of maternal age is less consistent, and possibly in the reverse.[103] As telomere length has been linked to both cardiovascular risk and longevity, this paternal effect on offspring telomere length has implications for human longevity, showing that some determinants may be decided even before we are conceived.

GETTING IN EARLY – PREVENTING PREMATURE SENESCENCE IN FOETAL LIFE:

We might imagine that cellular senescence and telomere shortening begin in earnest once we are older, but these processes are already taking place in foetal life. This can be subject to external environmental influences as well as genetic determinants. In-utero stress can lead to more rapid telomere shortening.[104] This cellular senescence can be further exacerbated with poor conditions after birth. 'Optimising' conditions in this early period of life is directed at maximising our biological resilience and reserve into our maturity. This needs to be aimed at all aspects of our biological and psychological development.

Let's begin with the placenta. Levels of the telomerase enzyme in the placental tissues are particularly high in early pregnancy, possibly to compensate for its very rapid growth before strong vascular supply is established. After the first trimester, telomerase levels fall progressively in normal pregnancies. Placental telomerase levels are significantly lower in pregnancies complicated by poor foetal growth or high maternal blood pressure (pre-eclampsia).[105, 106] The telomeres are also shorter.[107] These effects may be due to increases in oxidative stress and DNA damage in the placenta.[108]

Placental telomerase levels are also low or even absent in spontaneous miscarriages or foetal death.[109] And when twins are born with significantly discordant weights, the growth-retarded twin has been shown to have lower placental telomeres levels than the normal-weight twin.[110] We don't know how this might be translating in foetal tissues. But there is no doubt that poor foetal growth has consequences for adverse metabolic programming and the risk of NCDs, as discussed in earlier chapters. There has been increasing speculation that the cellular senescence contributing to the onset of these diseases may be initiated in this early period.

Low-birth-weight animals also show changes of accelerated senescence in kidneys and hearts, particularly after rapid catch-up growth after birth.[111] Similar effects in the pancreatic islet cells (Chapter 5) may predispose to diabetes.[112] But timing is important, because while protein restriction *before* birth decreases telomeres' length, restriction *after* birth increases it, as seen in adults (above).[113]

Senescent cells affect their healthy neighbour cells by secreting cytokines, proteins, and other factors to advance the ageing process in organs and tissues. Our best hope to mitigate this is to promote optimal nutrition and to reduce physical, chemical and psychological stress, particularly during pregnancy and the early years. We can also investigate factors that may reduce oxidative stress. Avoiding toxic exposures such as cigarette smoking, for example, which causes oxidative stress and cell senescence, even in utero[114] and in infancy.[115] In our own studies we are investigating the epigenetic effects of maternal fish-oil supplementation, and we are also now studying the effects that might have on the newborn telomeres.

EARLY NUTRITION, BREASTFEEDING AND HEALTHY AGEING:
Throughout pregnancy, optimal nutrition means a nutritionally balanced diet, and maintaining a steady, healthy pattern of weight gain. Gaining too much or too little weight in pregnancy may increase the risk of obesity, heart disease and diabetes and their many complications in later life, especially when growth restriction in pregnancy is followed by rapid 'catch up' and over-nutrition after birth (Chapter 4).

Breastfed infants generally show healthier growth patterns and, in some studies, are less predisposed to type 2 diabetes, with lower fasting insulin concentrations and insulin-like growth factor 1 (IGF-1) than their formula-fed counterparts; but the data on this are mixed. Breastfeeding appears to also confer favourable effects on blood cholesterol, although the long-term effects on cardiovascular protection are less consistent.[116-118]

Breastfeeding has also been linked to better cognitive outcomes in a number of studies.[119] A meta-analysis of *observational* studies of term infants found no overall effects,[120] but data from a large clinical trial (the PROBIT study) showed that the *interventional* study to improve breastfeeding rates through active promotion was associated with a 5.9 point increase in IQ.[121] In this context the levels of fats, namely n-3 PUFA levels, are important for brain development. Indeed dietary supplementation of docosahexaenoic acid (DHA) appears to enhance cognitive development in breastfed and formula-fed infants.[122] In our own earlier studies, we showed that DHA-rich fish oil in pregnancy improved n-3 PUFA levels in cord blood and in breast milk, with improved cognitive development in infants.

Complementary feeding and early nutrition are both important, and are discussed in Chapters 4 and 10.

• • •

The most challenging question we face is: just how much of the NCD risk seen in ageing can we attribute to early life? It is difficult to answer this exactly. An 'unhealthy' start to life will reduce biological reserve, but this is then overlaid by maladaptive responses, and then by ongoing unhealthy behaviours. Some of this is impossible to regain. If we have fewer nephrons in our kidneys, if we have fewer islet cells in our pancreas, if our peak bone density is low, and if we have fewer neuronal synapses in our brain, our 'reserve' will be lower. We will cope less well with the age-related decline in all of these tissues. And we will be less resilient to challenges. Then, superimpose the added maladaptive metabolic responses that underlie conditions such as obesity and metabolic syndrome. Once these become established they are very hard to change. If it may be possible to prevent much of this spiral with a 'healthy start', is there any question that we should try? Our relationship with food is one of our most important relationships in this equation. It holds the keys to maximising our potential and our longevity. Healthy ageing is about doing what we can to maximise our potential, and that must clearly start as early as possible.

12

A new world of opportunity

At its core, this book is as much about the health of our societies, of our economies, of our environment and of our human spirit as it is about the biology of our health. On the larger scale, the health crisis and its wide disparities are measures of the steepening gradient of inequity and social injustice. Human disease is set to cripple the global economy and add further to the pressures on global resources. We have had so many clear warnings of the absolute necessity to change, both what we are doing and how we are doing it. It is vitally important that we see these, not necessarily as impending doom, but as a vigorous call to action.

In the preceding chapters we have seen how the health of our societies and the health of individuals are inextricably inter- twined on many levels. We have seen how many of the 'common risk factors' for disease (diet, inactivity, smoking, pollutants, time indoors, and declining microbial biodiversity) are inherently driven by wider social, cultural and economic factors, and the quality of our built environment and our natural environment.[1] The WHO report on *Social Determinants of Health: The Solid Facts* summa- rises the clear relationship between life expectancy and social standing – with an increasing burden of disease and premature death as we go down the social ladder.[2] Social exclusion, lack of self-determination, including lack of control in the working environment, unemployment, job insecurity and stress, all increase illness and premature death. Higher use of alcohol, cigarettes and drugs is also a response to social stresses and social breakdown, further adding to the social gradient in health. Similarly, poor food quality and unhealthy eating patterns are much more likely in the socially disadvantaged. As we have seen, all of these factors contribute to poor maternal health and family stress and adversely influence the physical and psychological development of the next generation.

Inherent to this is the relationship between the health of individuals and the 'health' of their community. People with more social connections are physically and mentally healthier, and they live longer,[3] showing the clear benefits to the individual. At the same time, social networks and participation also promote a better community, showing the collective benefits for society. There are now good studies linking social isolation to poverty and economic disadvantage, although this can vary with context.[4] While this is a bi-directional relationship (cause and effect both operating), it nonetheless highlights the interplay between the health of individuals and collective humanity.

Trying to address only the superficial 'causes' of NCDs (i.e. bad nutrition, smoking and inactivity) without addressing the wider, so-called *'causes of the causes'* (i.e. the social, cultural and economic determinants of health)[5] will be certain to fail. Essentially, to use a medical analogy, it is like treating each separate symptom while completely ignoring the underlying disease process.

THE HEALTH EFFECTS OF INEQUALITY:
I take much inspiration from author, strategist and noted commentator on world events, Mark Haynes Daniell. Also a dear friend, Mark is author of the *World of Risk – A New Approach to Global Strategy and Leadership*. He underscores the importance of effective leadership, clear vision and strategic direction in achieving the

societal transformation we so desperately need. Ultimately though, he reminds us that our greatest hope *'abides in the potential force for change which resides in each and every one of us as individuals'*.[6]

Bringing this back to matters of health, there is no question that a sense of personal empowerment and connection to community bring better mental health and physical health. In all of this, a sense of greater purpose is a critical ingredient for health at every age, and builds the foundations of our longevity.

The story of health disparities, with their social and economic causes, begins very early in life. Nowhere is this clearer than in Australia, where the estimated life expectancy of Aboriginal and Torres Strait Islander peoples was approximately seventeen years lower than non-Indigenous Australians at the turn of the new century (1999–2001).[7] Babies born to Indigenous Australians are twice as likely to be born with a low birth weight than non-Indigenous babies. The infant mortality rate is also twice as high, reflecting the greater adversity of the early environment. This more difficult 'start to life' sets the scene for higher rates of death and disease at all ages. These inequalities in health are completely avoidable[8] and *cannot, and should not, be separated* from the wide disparities in education, housing conditions and other socioeconomic indicators. Indigenous youth are more likely to show psychological distress and teenage pregnancy, live in overcrowded conditions and are fifteen times more likely to be in juvenile justice supervision or prison. Without addressing these underlying social determinants of health, the health divide cannot be overcome. This has lead to an all-of-government 'Close the Gap' initiative, built strongly around engagement with the Aboriginal and Torres Strait Islander people. Progress has been slow, with still around 10 years' reduced Indigenous life expectancy.[9, 10] Many of the targets required to 'close the gap' are still falling well short; however, a decline in Indigenous child mortality is some cause for optimism. Between 1998 and 2012, Indigenous child mortality has declined by 32 per cent, outpacing the decline in non-Indigenous child mortality.[11] While the benefits of improving child health and education may not been seen for many years, their potential for long-term impact should not be underestimated. There is still much to be done.

Thirty years ago similar health inequalities could be seen for the Indigenous peoples of Canada, New Zealand and the United States of America.[12] But those regions have since shown major progress in improving health and life expectancy by addressing the socio-economic divide. Engagement and empowerment are key elements of success in these regions, while Australia has fallen behind on all of these indicators.

These examples show how the health of any population is an important measure of the effectiveness of economic and social policies, and whether or not they are benefiting the population and adequately addressing all of its members. This is the central dogma of the WHO Commission on Social Determinants of Health: 'policies that harm human health need to be identified and, where possible, changed. From this perspective, globalisation and markets are good or bad in so far as the way they are operated affects health'. [13]

With all of our challenges, this is also an age of unparalleled opportunity and there is great cause for optimism. We have never been better equipped. The technology to achieve transformation is building exponentially. And the level of mass global connectivity through social media and other networks is unparalleled.

THE AGE OF CONVERGENCE:

We are slowly becoming convinced that all systems on Earth are inter-connected and that many of the problems we face today are the consequences of the decisions and philosophies of yesterday. As a result, there an emerging emphasis and renewed focus on 'sustainability'.

This requires integration of all of our key social, economic, or environmental needs into a single model, even those in apparent conflict. This is my purpose in finding common and shared risk factors for almost all NCDs, and for linking this with the health of our global society, the health of our global economy, and the health of our global environment. In this we may find a convergence of purpose, and a convergence of vision for implementing an integrated plan of action.

When different sectors come together on common ground, the results can be powerful. One such good example, with simultaneous benefits for child health and for local economies and agriculture, comes from the school meal program in Brazil. While there is no short-term financial incentive to provide fresh, healthy food to school children, by setting a goal that 30 per cent of school food must be locally produced by local farmers, the government provided the local communities with strong agricultural incentives which had immediate local benefits. This has been highly successful, has fed more than 45 million children and invested around $500 million in family farmers in 2010. In collaboration with the United Nations Food and Agricultural Organization, similar programs are now being established in Africa.[14]

This is a wonderful example of the kind of creative and collaborative approach that is needed. The challenge is now for us to find the win–win element in every situation. To do that we need to be speaking the same language and have a 'collective awareness' and a more 'convergent' state of mind.

A number of elements need to be balanced in creating sustainable social systems. These need to include addressing household needs, empowerment, engagement, social mobility, efficient use of labour, equity, biodiversity, services and infrastructure, industrial and agricultural growth, and preservation of culture and the environment. In the first instance, many of the existing programs centre on achieving these goals through improving education, empowering women, building infrastructure, establishing strong regard for social justice, and ensuring wide local engagement and ownership. In this, there needs to be clear understanding of the collective benefits, and how these flow on to the health and prosperity of the community.

RECALIBRATION:

Humans are adaptable. We can live with a little less salt, less fat and less sugar in our food. Even eat a bit less. We could get used to it, and find we are much healthier. We might manage a bit more physical activity, and even enjoy it. We could learn to slow down, relax and actually find that we are less stressed. We might find ways of living that reduce the level of pollution that we produce, both as individuals and as a collective, and find we enjoy our relationship with the environment more. Those of us with a good financial position might even be content to have a bit less to allow others to have a bit more, without going hungry or becoming destitute. This might even make us happier.

We need to recalibrate. For our own sake. It is the only logical way forward.

For the new generation, less healthy lifestyle patterns may adversely influence both early biological development and the 'normal' early lifestyle behaviours that follow – making this harder to 'recalibrate'. For animals in captivity, it can take several generations to return to good health under more natural conditions. We would do well to act sooner rather than later.

THE ROLE OF GOVERNMENT AND BUSINESS:

The corporate sector is critically important for social change. We should look to the business leaders to lead greater corporate responsibility and engagement in social and economic initiatives to address health and inequity.

In his clarion call to businesses, former Secretary-General of the United Nations, Kofi Annan, had this to say: '*We cannot wait for governments to do it all...Business, labour and civil society organizations have skills and resources that are vital in helping to build a more robust global community'.*

Fast food is cheap. So much more of what we eat is processed or factory prepared. Fresh food can be expensive and may have been transported large distance across the globe, adding to our 'carbon foot-print' and environmental degradation. How do we recalibrate our food? How do we shift our demand and our supply to produce healthier food that we can still enjoy?

Like all other businesses, food companies operate to make a profit in a free market, the system of our collective creation. They are not inherently evil. But there could be more checks and balances. A collective agreement is needed for mutual benefit, because human health is at stake. Even if this is driven by the need to address the economic burden of NCDs.

There have been many government initiatives to engage and influence the food industry, but most of these have failed to effect meaningful changes. The Australian government established a Food and Health Dialogue in 2009, to engage the food industry and other key stakeholders. However, a recent independent review of progress revealed that few goals have been set and none have been achieved.[15] This report identified the limitations of voluntary industry codes, and the need to better control conflicts of interest. And the authors pointed out a great irony – that the food industry carefully adhere to stringent regulations to prevent 'acute poisoning', yet there are no regulations and virtually no regard for the 'chronic food poisoning' inflicted by their high-fat, high-sugar, high-salt products.

Some large multi-national companies have committed to making small incremental changes in product content, and so have their competitors. But there is a long way to go. More pressure is needed on every front. These sorts of wider initiatives will be more effective on a larger scale – more consistent, and more sustainable than only relying on individuals to regulate their own behaviour.

This underscores the importance of education, especially around the importance of early intervention for the greatest effect.

Then we come to environmental toxins. Can we imagine a hypothetical situation where a very large and wealthy entity might be freely allowed, even encouraged, to openly and *deliberately poison* billions of people, to *knowingly* cause immeasurable harm, death and disease? It seems preposterous that such a hypothetical entity could *ever* be allowed to make immense profits for decades, crippling economies, without strong action from its victims or the world's governments. Yet this happens every day – and has been happening for years. Cigarettes. Pure exploitation for profit. How can governments and our leaders continue to allow this, with such ineffectual legislation? How have the rest of us tolerated this for so long?

Considerable progress has been made. Attitudes have shifted. But there is a long way to go. There needs to be a massive recalibration of what is acceptable in our world and what is not.

LIVING FOR FUTURE GENERATIONS:

What we need is, very clearly, not always what we want. Our current pandemic of obesity and NCD is only one of the consequences of this. But we all bear the costs – either directly or indirectly. And we all need to take responsibility for changing this.

The Native American Iroquois (Haudenosaunee) Nations measured their deliberations against what was just and right for the *future* of their nation. And they gave thought, not just to the next generation, but through to the seventh generation hence.

'Look and listen for the welfare of the whole people and have always in view not only the present but also the coming generations, even those whose faces are yet beneath the surface of the ground – the unborn of the future Nation.'[16]

Truly 'looking ahead' does provide an important set of codes and values that can be used in the face of complexity and competing interests.

THE SHAPE OF THINGS TO COME?

As we come to the end of the journey through these pages, it is a moment to reflect on where we have come from and where we are going as we begin to anticipate the health challenges of the new century.

If we look back across the twentieth century we can see enormous gains in life expectancy achieved in the last 100 years, largely due to reduced infectious diseases and, in particular, improved maternal and child health. But there are still wide discrepancies, both within and between societies. And poor conditions in early life remain both a major element in these health inequalities, and a critical factor in overcoming them.

As we look forward into the twenty-first century, we are facing very different health challenges. Entering a new age of technology, we may have once imagined more leisure and better health – but instead we have become fatter, busier, sicker and more stressed. And this is occurring progressively earlier in life, with rising rates of immune diseases, obesity and metabolic diseases, and mental ill-health in very young people. These new epidemics of early-onset NCDs will have major implications for the global disease burden as this generation reaches maturity.

Action is needed. But there will be no quick solutions.

The momentum of our current health crisis has built over decades, and it is likely take just as long to turn it around. This means that we must take a 'long view'.

A better understanding of our biology has revealed the critical importance of a 'developmental approach' to health, beginning from the first moments of life, and then through the many windows

271

of opportunity beyond. It has also taught us that the health of parents at the time of conception is vital in the health of the next generation. In this knowledge it is imperative that we promote health, education and wellbeing in the *next* generation of parents before they reach maturity – the children and youth of today. Fundamental to this are the gender inequities in many regions that must be overcome, so that young girls and women, who carry the health of the generations to come, have every opportunity to do so.

Our biology also reveals our resilience and our inherent adaptability to the environment. Paradoxically, this lies at the very heart of many of our maladaptive responses to the very rapid modern environmental transition. But the same plasticity gives us great cause for optimism – that the keys to reversing our problems lie in harnessing the same biological pathways to our advantage.

We already have a broad understanding of many of the lifestyle and environmental risk factors, shared by so many NCDs. And we already have a broad sense of what we need to do to address these. We now just need to understand *how* best to do it. This will require a deeper understanding of the biology, including how the multitude of risk factors interact with genetic risk, the most effective interventions and how to implement them.

We may anticipate this will lead to both generic strategies at a population level as well as individualised strategies to prevent disease, according to specific genetic or environmental-risk scenarios. Importantly, this will require a far more collaborative, cross-disciplinary approach, which builds on the understanding that shared risk factors need shared solutions – which will bring benefits not only to our physical and mental health, but also to our economic and societal health. Unifying concepts and common goals will help us make sense of complexity and help us simplify the solutions.

That said, it is also clear that the origins of our modern-health crisis go far beyond biology. And so must the solutions. This means that we require an integrated global approach that recognises the wider social, cultural and economic determinants of health, as well as the complex interrelationship between human health and all of our other major global challenges. Einstein is noted for saying that *'No problem can be solved from the same level of consciousness that created it'*. We might anticipate that addressing our complex problems from a higher integrated perspective might provide simpler solutions than might be otherwise anticipated. This does not mean the solutions will be easy, but it could mean that a more overarching approach might address many of our problems collectively.

In searching for some final perspectives, I was drawn back to my own words, which previously described the 'way forward' in the context of the *Allergy Epidemic* and which seem equally pertinent to our many other health challenges:

> *'We need to be open to new ideas. New approaches. But we also need to find solutions that will not create more problems. Strong scientific rigour is needed to continue to pursue new approaches as they arise, as well as the many existing avenues of investigation. This is a global problem and a global effort to combat this is now well underway. All this will take time, but there is cause for great optimism and anticipation'.*

And I remain optimistic. We have never been better equipped to deal with these global challenges. We may be facing a crisis of our own making but we now have the resources and the technology to overcome it. This calls for a creative collaborative vision, and collective responsibility at all levels – individuals, industry and governments. Transformation is best achieved by aiming high and by not being afraid to do so.

In closing, I also return to the reasons that I set out on this journey in the first place, and my purpose in telling this story of health, through this book. And I reflect on how much has happened since I travelled the dusty roads of India in 2001, for the very first international meeting to focus on early-life programming. It was the birth of the 'development orgins' as a new field of medicine. A concept that is intuitive on so many levels, and strong in many traditional cultures, but generally overlooked by orthodox medicine at that time. In the fifteen years since, there has been an enormous extension of this knowledge, and a deeper understanding of the breadth and extent of long-term latent effects on our health and longevity. The discovery of 'epigenetics' has provided a clear biological basis for these effects on biological programming.

Scientists are not always good at 'getting the message out'. And I wanted to share these discoveries and their significance with the people who most need to know, and who can best do something about it.

I wanted to tell this story, not just because it is important for understanding our health, but because it is important for understanding our future.

Many of today's problems are the consequence of yesterday's decisions. And this means that the solutions of tomorrow must begin today. The health of our future, and the world our children will inherit, will depend on the choices we make together now.

We have crossed the threshold and now the journey must begin.

Postscript

Slow down and enjoy the ride.

A very good reason that we haven't managed the change needed in our societies is that we are just too busy. We are also too busy to eat well, too busy to exercise, and often too busy to do anything for our community. We hardly have time to look after ourselves. The stress of it all drives many to smoking, alcohol and other balms to escape the growing chaos. I don't know many people who don't say they are too busy. It has become our modern mantra.

This stress is bad for our health. So is the lack of exercise. So is the 'fast' food or the 'comfort' food that we are more inclined to eat. We eat on the run. Or we eat in front of the TV. We don't sleep as well. Not to mention our sex drive. And we feel more depressed about how unhealthy we feel. We lack energy and probably drink too many caffeinated beverages, and crave more sugar and fatty foods to keep us going. We are short-tempered, reactive and don't really contemplate problems or situations more deeply. And this just adds to the stressful situations that we find ourselves in.

There are other less-obvious aspects of this that are bad for our health. The speed of life is also part of the reason for our disconnection with our community and with friends, and our disconnection from nature. We are too busy to see our family and friends as much as we might like. And too busy for a walk in the forest. It is quicker and easier to stay at home, and watch TV.

That is all we have the energy for. We know that isolation and disconnection from community are specific risk factors for poor health and disease. And we are just beginning to learn that our disconnection from nature is also having specific and unexpected effects on our mental and physical health. We should be concerned that our children are also losing this important relationship with nature. Instead we sit in traffic or we speed about in a frenzy. We know this is madness, but we seem to do it anyway.

One day, after a rush to get from work to meet my husband at the movies, I almost had a serious accident. It seemed like life was warning me to slow down or there would be serious consequences. As I was catching my breath and trying to 'slow down' and calm my nerves during the movie trailers the following advertisement came onto the screen and had me transfixed. It was part of a major campaign by my state government in Western Australia, from the Office of Road Safety. Although the campaign was addressing road safety, the general messages for our health and life behaviour were so good, I want to share them here.

> *If life's a race, where's the finish? And who are we competing with? Is there a prize for first place? Or do we just reach the end a little quicker?*
>
> *We're only just skimming the surface of life. Humans just aren't designed to go that fast. Sooner or later we crash.*
>
> *To cope with this speeding life, our bodies release chemicals that activate our adrenal glands, increase our heart rate and raise our blood pressure. These responses cause us to grind our teeth, sleep poorly, crave fatty and sugary food, get headaches, feel stressed, get sick more often and lose our sex drive.*
>
> *And in the long term we're in the fast lane to heart disease, sexual dysfunction, allergies, diabetes, bowel conditions, depression, anxiety, muscular pain... and a load of other stuff with names too long to remember.*

But when we slow down, we discover that life has a natural pace. And it's good! We slot into a groove that's always been there. Life becomes richer. More pleasurable. And more fulfilling. We may do fewer things, but what we do, we do well.

We breathe! When was the last time you actually took time to breathe? Not just the shallow ticking over of your respiratory system. But to really breathe. Taking a long slow breath in to its comfortable conclusion. Then letting it all out. And doing it again... and again... and again, until you are flooded with calm.

Imagine life lived in this zone. So why haven't we slowed down before? If you're worried life will overtake you, you're wrong. Life is where you are and what you're doing right now... and now... and now.

Let's say two cars take the same route. One speeds. The other relaxes and follows the speed limit. The speeder spends his trip tense – at risk of being caught by cops and cameras. The other is free to enjoy the ride.

Sure the speeder gets to the lights first...only to be caught up by the guy going the right pace.

There's one place where we can all enjoy life at the right pace. Treat your vehicle as a sanctuary. And shut your door on the world of speed. When you pull away, leave speed behind. If we fight traffic, we only ever lose. Besides, driving can be a pleasure, if we don't treat life like a race. So slow down, and enjoy the ride.[1]

Most of us want to, *need* to, step off the treadmill. Even for a few moments each day. Time to reflect. Time to connect to ourselves. Close down our emails, put down our tablets. And just slow down. And breathe.

We need to remember that the frenzied pace of life is of our 'collective' creation. And that we all contribute to it in our own way. That means the only way we can slow down the collective is by changing our individual behaviour. For our own sake, but also for the wider collective benefit.

This is an important, probably unrecognised, part of the solutions to our many global problems. It goes to our personal health, the health of our community, the depth of our connections, and the sense of higher purpose and collective responsibility that is needed to make societal change. We can all play our part in slowing down the pace of life. Starting with ourselves.

Glossary

adipocytes: the fat-storing cells that make up body fat (adipose tissue).

adipostat: the physiological processes that balance metabolism and body fat.

allergic disease: conditions such as asthma, and allergic rhinitis, food allergy and eczema, which result from inappropriate immune responses to the environment.

anabolic: refers to the 'building' aspects of metabolism which are active during tissue growth and maintenance.

antioxidants: substances that protect cells from damage caused by unstable products of metabolism known as free radicals.

antigens: fragments of any (environmental or self) protein.

antigen presenting cells (APC): the cells that digests proteins into antigens to be processed by the immune system.

ART (IVF): Assisted Reproductive Technology, including in-vitro fertilisation.

atherosclerosis: thickening of arteries due to plaques formed by fat containing immune cells.

autoimmune: an immune reaction to our own tissues (self antigens).

autophagy: a cellular 'recycling' process that degrades and reuses unnecessary or dysfunctional components.

B cells: a major class of lymphocytes (immune cells) that produce antibodies.

brain-derived neurotrophic factor (BDNF): a nerve growth factor, important for learning memory and many higher brain functions.

BMD: bone mineral density.

bifidobacteria: species of normal healthy gut bacteria, also used in probiotic products.

biphenyl compounds or PCBs (polychlorinated biphenyl): synthetic compounds used in paints, electrical wiring, cements, plastics and many modern materials.

BPA (bisphenol-A): synthetically produced chemical that has been used to make certain plastics.

brain axon: nerve fibre that conducts electrical impulses.

BRCA1 and BRCA2 genes: mutations of these genes is associated with breast cancer risk.

catabolic: metabolic processes involved in the 'breakdown' of molecules to release energy.

cortisol: the main hormone regulating our stress responses.

C-reactive protein or CRP: a protein produced by a liver in response to inflammation anywhere in the body.

cytokines: signals produced by many cells to influence the growth and activity of other cells and tissues.

endocrine disrupter: environmental chemicals that may mimic the effects of some hormones.

epigenetics: the complex mechanisms that control the patterns of gene expression in each individual cell (Chapter 11).

DOHaD: developmental origins of health and disease.

dioxins: toxic chemical compounds released during burning (forest fires or the burning of commercial waste) and other industrial activities.

dual-energy x-ray absorptiometry ('DXA'): a radiographic measure of bone mineral density.

endotoxin: a component of certain bacteria, also known as lipopolysaccharide.

ghrelin: the 'hunger hormone' released from our stomach when our stomach is empty.

glucocorticoid receptor (GR): the main receptor for the cortisol stress hormone.

Helicobacter pylori **(H. Pylori):** a bacteria that naturally occurs in the stomach, and is associated with ulcers in some circumstances.

hippocampus: a part of the brain associated with memory, learning and behaviour.

homeostasis: mechanisms that regulate and stabilise various body functions in an optimal range.

hygiene hypothesis: cleaner environments and declining biodiversity may be increasing immune and inflammatory diseases.

hypertension: high blood pressure.

hypothalamic-pituitary-adrenal (HPA) axis: the group of organs involved in hormonal regulation of cortisol levels and stress responses.

insulin: the hormone produced in the pancreas that regulates blood-sugar levels.

insulin-like growth factor-1 (IGF-1): an 'anabolic' growth factor related to insulin.

interleukin-6 and interleukin-1 beta: cytokines that promote inflammation.

ketones: substances produced when the body breaks down fat for energy.

lactobacillus: species of normal healthy gut bacteria, also used in probiotic products.

leptin: the 'satiety' hormone which regulates appetite and fat metabolism.

lymphocytes: a major class of immune cells, which includes B cells, T cells and other lymphocytes.

macrophages: a scavenger immune cell (a form of 'antigen-presenting cell' as above).

metabolic/metabolism: the processes involved in energy utilisation and storage for body functions.

metabolic syndrome: abnormalities of metabolism generally associated with obesity, high blood pressure, blood sugar and cholesterol.

microbiota: the trillions of bacteria that naturally live in and on the human body.

microbiome: the collective genetic material of the microbiota (although the terms are often used interchangeably).

microglial cells: the immune cells of the brain (another class of antigen-presenting cell).

mismatch theory (of evolution): traits that evolved (over millennia) to be adaptive in a certain environment, are now 'mismatched' (not advantageous) in a new environment.

mismatch (developmental): physiological adaptations to the environment in early life (e.g. under-nutrition) do not match subsequent conditions (e.g. over-nutrition), increasing the risk of disease.

nephrons: the basic structural and functional units of the kidney.

network theory: the study of complex interacting systems.

neurons: nerve cells (also see synapses).

NCDs: 'non-communicable diseases'.

nucleotide: the individual building blocks of DNA (and RNA).

oncogenes: a gene that contributes to transforming a normal cell into a cancer cell when mutated or expressed at abnormally-high levels.

organochlorine pesticides: synthetic organic pesticides that resist metabolism and accumulate in fatty tissues.

osteoporosis: a condition of decrease in bone mass and density ('porous bones') and increased risk of fracture.

oxidative stress: when production of potentially damaging metabolic by-products (free radicals) exceeds the capacity to detoxify.

phthalates: chemicals used in plastics to make them more flexible and harder to break.

polymerase (DNA polymerase): enzymes that assemble nucleotides to create DNA molecules.

PUFA: omega-3 polyunsaturated fatty acids in fish oils.

prebiotics: a form of soluble dietary fibre (oligosaccharides) which induce the growth and/or activity of healthy gut microbiota.

probiotics: dietary supplement containing bacteria (such as lactobacilli and bifidobacteria) that are shown to restore beneficial bacteria and provide health benefits.

RNA (ribonucleic acid): messenger RNA is 'transcribed' from DNA and transported out of the cell nucleus where it is 'translated' to make specific proteins.

senescent cells: when cells reach their replication limit and no longer divide; associated with cellular ageing and can be induced by certain toxins and irradiation.

short chain fatty acids (SCFA): fermentation products of dietary fibre (prebiotics) with anti-inflammatory properties.

synapses: the structures that allow nerve cells to pass electrical or chemical signal to one another or to other cells.

telomerase enzyme: the enzyme which adds nucleic acids to DNA to elongate telomeres.

telomeres: the cap at the end of each chromosome which protects DNA from erosion during cell division, but shortens progressively with age.

triglycerides: a form of fat (lipid) for storing unused calories; increased levels in blood are used as a measure of cardiovascular risk.

T cells: a major class of lymphocyte, involved in generating immune 'memory' to previously encountered antigens.

UWA: University of Western Australia.

WUN: World Universities Network.

Appendix

On the practical side – what is currently recommended in early life?

A group of my colleagues recently got together to develop some practical recommendations and advice to families and parents. With their permission, the following information is taken directly from those recommendations. They gave particular focus to early-life nutrition, as one of the most important and easily modifiable environmental factors during early life. Although these are directed at families in Australia and New Zealand, the advice is also relevant to many other regions. These are evidence-based recommendations to maximise nutritional status before and during pregnancy, as well as during infancy and early childhood, when the foundations of heath are created.

> This work is included courtesy of authors Peter Davies, John Funder, Debbie Palmer, John Sinn, Mark Vickers, and Clare Wall, for their work on 'Early life nutrition. The opportunity to influence long-term health'. 2014[1]

RECOMMENDATIONS

'Pre-pregnancy, pregnancy, infancy and early childhood represent key windows of opportunity for parents to adopt lifestyle and nutritional strategies that can improve foetal and childhood development and lower the risks of

their children developing allergic and metabolic disease in later life. Practical, evidence-based recommendations can support parents and healthcare professionals to maximise foetal and childhood development during the period in which the key foundations of future heath are created'.[2]

1. PRE-CONCEPTION AND PREGNANCY

Reduce excess weight prior to conception: Obesity is now one of the most common and important risk factors for infertility and adverse pregnancy outcomes.[3] It reduces the likelihood of becoming pregnant, increases maternal complications during pregnancy[4] and is also associated with an increased risk of obesity in the offspring in later life.[5, 6]

- Women should aim to achieve and maintain a healthy body weight prior to becoming pregnant.
- Women who are obese (BMI of 30 kg/m^2 or more) should be advised and encouraged to safely reduce their weight before becoming pregnant. This may require specific education about nutrition and physical activity strategies from appropriate specialists.
- Losing 5 to 10 per cent of body weight can have significant health benefits for the woman and may also increase the chances of becoming pregnant.[7]
- Women should be supported in their weight control endeavours by their partner.

Maintain healthy weight and monitor weight gain during pregnancy: Excessive weight gain and obesity during pregnancy can have an adverse impact on a woman's health and the long-term health of her child – in particular an increased risk of obesity, coronary heart disease and type 2 diabetes in later life.[8, 9]

- Steady weight gain during pregnancy is normal and important for the health of the mother and baby. However, it is important that women who are pregnant do not gain too much weight.

Weight gain should be discussed by healthcare professionals and monitored regularly during antenatal care.

- Women should be advised about the appropriate amount of weight to gain during each stage of pregnancy. *Recommendations are outlined in Table 1.*

- Just as overconsumption can be damaging to the developing foetus, a lack of nutrients is also problematic. As such, dieting or weight-loss programs during pregnancy are not recommended as they may compromise the health of the unborn child.[10]

- Women may notice changes in appetite and taste preferences during pregnancy, and these generally settle down over time. Women should satisfy their appetite but continue to eat a healthy diet and monitor weight gain so that it is not excessive *(See Table 1)*.

Table 1. Recommended weight gain during pregnancy.[11]

Pre-pregnancy body mass index	Recommended total weight gain
Less than 18.5 kg/m^2	12.5 to 18 kg
18.5 to 24.9 kg/m^2	11.5 to 16 kg
25 to 29.9 kg/m^2	7 to 11.5 kg
More than 30 kg/m^2	5 to 9 kg

Maintain a healthy diet and lifestyle prior to and during pregnancy: Couples planning a pregnancy should implement and maintain a pattern of healthy eating and physical activity in order to increase the likelihood of pregnancy and, in the case of the male, improve sperm quality.

- It may be appropriate to assess the presence of any nutritional deficiencies in women planning a pregnancy. Both Australia and New Zealand have evidence-based healthy-eating guidelines about the type and amount of foods that are recommended.

- Physical activity is also important for maintaining a healthy lifestyle prior to and during pregnancy. Women should be encouraged to do at least thirty minutes of moderate intensity physical activity on most if not all days of the week (i.e. a target of at least 150 minutes per week).
- A woman should stop smoking before attempting to become pregnant. Paternal smoking prior to conception should also be discontinued as it has been associated with damage to sperm DNA and an increased risk of malignancy in offspring.[12]

Stop smoking and alcohol use during pregnancy: Cigarette smoking (including passive smoking) and alcohol consumption during pregnancy can have serious health consequences for the women and her unborn child and should be avoided.

- Cigarette smoking increases the risk of ectopic pregnancy, miscarriage, premature labour, low birth weight and sudden unexpected death in infancy.[13]
- Drinking alcohol during pregnancy has been associated with miscarriage, low birth weight and intellectual impairment (known as foetal alcohol syndrome). Women should be advised that there is no safe level of alcohol consumption during a pregnancy.[14]

Optimise control of existing health conditions: Existing health conditions should be controlled prior to and during pregnancy in order to reduce the risk of complications during pregnancy.

- Women with chronic diseases such as asthma, diabetes and chronic kidney disease have an increased risk of complications during pregnancy.
- Women with diabetes who are planning a pregnancy should be informed that establishing good glycaemic control before conception and maintaining good glycaemic control during pregnancy has important health benefits, including a reduced

risk of miscarriage, congenital malformation, stillbirth and neonatal death.[15]

Take folic acid supplementation: Maternal folic acid deficiency has been linked to the development of neural tube defects (NTD) such as spina bifida in the offspring.

- It is recommended that folic acid supplementation is taken for a minimum of one month before conception to assist with the prevention of NTD.[16]
- Folic acid supplementation should be continued for the first trimester of pregnancy. There is no evidence that continued folic acid supplementation after the first trimester is beneficial.
- The recommended dose of folic acid supplementation during pre-conception is at least 0.4 mg daily. Where there is an increased risk of NTD (e.g. in women who are obese, those using anticonvulsant medication, those with existing type 2 diabetes, those with a previous history of a child with NTD or a family history of NTD), a 5 mg daily dose of folic acid is recommended.[17]

Take iodine supplementation: Severe maternal iodine deficiency has been linked to pregnancy loss and impaired mental and physical development in the foetus. Mild-to-moderate iodine deficiency during pregnancy adversely affects infant thyroid function and may affect mental development.[18]

- Iodine supplementation (150 mcg daily) is recommended before a woman becomes pregnant to help optimise foetal development and pregnancy, as well as for the duration of pregnancy and during breastfeeding.[19]
- Women with diagnosed pre-existing thyroid conditions should seek advice from their doctor before taking any iodine supplements.

- In addition, women who consume daily seaweed soup should seek advice before taking iodine supplementation, as it may lead to an excess of iodine and subsequently impact on thyroid function.

Maintain recommended dietary intake of omega-3 fatty acids: Omega-3 fatty acids are important for foetal development, including brain development.[20]

- Australian and New Zealand dietary guidelines recommend two to three servings of oily fish (e.g. salmon, tuna) per week to provide adequate amounts of omega-3 fatty acids.
- Most fish in Australia and New Zealand are low in mercury, but this varies depending on the type of fish. The higher up the food chain, the more mercury the fish is likely to contain.
- Too much mercury can harm the developing nervous system, so it is recommended that pregnant women are aware of the mercury levels of different types of fish and how often to eat each type. [21]

Consume appropriate levels of other vitamins and minerals: Maternal micronutrient status plays an important role in pregnancy and birth outcomes. Maternal deficiencies in certain micronutrients may have an impact on foetal development and the subsequent health of the offspring. Above and beyond the intake of folic acid, iodine and omega-3 fatty acids, pregnant women should maintain adequate levels of vitamins and minerals including the following.

Calcium: Calcium is required for the normal development and maintenance of the skeleton.

- Calcium supplementation is recommended for women who avoid dairy in their usual diet and do not consume alternative high calcium food (e.g. calcium-enriched soy milk).

- The recommended dietary intake of calcium for pregnant women is 1300 mg per day for those aged under eighteen years and 1000 mg for those aged nineteen to fifty.[22]

Iron: Requirements increase during pregnancy in order to provide for the growing foetus and increased maternal blood volume.

- Maternal iron-deficiency anaemia has been shown to adversely affect foetal brain development, increasing the risk of poor cognitive as well as poor motor and behavioural development in the offspring. Children of iron-deficient mothers are also more likely to have low iron stores and be susceptible to iron deficiency.
- Iron status in pregnant women should be monitored, and iron deficiency treated with iron medications. All women should receive advice on dietary sources of iron and factors affecting iron absorption.[23]

Zinc: Essential for growth and neurobehavioural development of the foetus and its requirements increase during pregnancy.

- Maternal zinc deficiency has been associated with growth retardation and congenital abnormalities (including neural tube defects), low birth weight and premature delivery, as well as problems with neurobehavioural and immunological development in the foetus.[24]

Vitamin B12: Essential for cell function and neurological function, including neural-tube development.

- Maternal deficiency has been associated with impaired neurodevelopment in infants.
- Low plasma vitamin B12 status in early pregnancy has been associated with a significant elevation in insulin resistance in the offspring.[25]

- The recommended daily intake of vitamin B12 is 6 mcg/day. Vegetarians and vegans should be advised to eat foods that contain vitamin B12, such as milk and milk products, eggs and/or foods fortified with B12 (e.g. soy milk). They should also receive vitamin B12 supplementation during pregnancy and breastfeeding.[26]
- It is important to note that vitamin B12 in plant sources, such as seaweed and spirulina, does not translate to vitamin B12 activity in the human body.[27]

Vitamin D: Severe maternal deficiency has been associated with rickets in the offspring. Maternal vitamin D deficiency has also been associated with decreased foetal growth through its effect on maternal calcium homeostasis,[28] and may also affect bone mineralisation in adulthood at the time of peak bone mass.[29]

- Women who are at risk of vitamin D deficiency should be identified and advised about methods to increase vitamin D levels. This includes women with reduced sunlight skin exposure (e.g. women who wear a veil), those who use sunscreen on a regular basis, dark-skinned women and obese women.
- For women who are at increased risk of vitamin D deficiency, testing should be considered and supplementation instituted where needed.[30]

Increase protein and carbohydrate consumption and moderate fat intake: Increased consumption of protein and carbohydrate is required during pregnancy to promote foetal growth and development; however, fat intake – especially saturated fat – may need to be reduced.

- **Protein:** The requirement for protein increases during pregnancy to support foetal growth, especially in the third trimester. However, high protein diets (over 20 per cent of total energy) should be avoided as they may lead to increased birth weight.[31]

- **Carbohydrate:** Increased intake of carbohydrates is important in pregnancy to ensure adequate glucose supply for maternal brain metabolism, as well as for foetal development. Pregnant women should aim for eight-and-a-half serves of breads and cereals (preferably wholegrain) each day.[32]
- Pregnant and breastfeeding women should aim to reach the recommended fat intake level of 20 to 35 per cent of energy intake. This may mean a reduction in fat intake, especially saturated fat.

Consumption of allergenic foods: The developing foetus and newborn infant benefit from exposure to a wide range of nutritious foods consumed by the mother during pregnancy and breastfeeding.

- Avoiding allergenic foods such as peanuts, peanut products and tree nuts during pregnancy and breastfeeding with the aim of reducing allergy risk in the offspring is not necessary.[33] However, if a pregnant or breastfeeding woman has a food allergy, she should continue to avoid those foods for her own safety.

Prebiotics and probiotics: A healthy balance of micro-organisms in the gut is essential for healthy immune system development and metabolic regulation.[34, 35]

- The use of probiotic bacteria, commonly found in yoghurt, yeast and supplements, during pregnancy has been shown to have metabolic and immune system benefits.[36] A study found that combined dietary counselling and probiotics improved insulin sensitivity and glucose metabolism in healthy women.[37]
- There is emerging evidence that the use of prebiotics (naturally occurring dietary fibre which acts as food for probiotics)

during pregnancy may protect against allergy in the offspring long term. Changing the gut flora of the mother can change the microbiota of the infant. A systematic review found that prebiotic use in late pregnancy and early infancy was associated with a significant reduction in the development of eczema.[38] Further studies are required to determine the effects of prebiotics in pregnancy.

Women with additional needs: Pregnant women with the following conditions, or in the following situations, should talk with their doctor about nutrition during pregnancy and breastfeeding:

- morning sickness
- low gestational weight gain
- pre-existing diabetes or gestational diabetes
- smokers, illicit drug users, women who continue to consume alcohol
- teenage mothers.

2. NEWBORN AND INFANT – FIRST SIX MONTHS

Breastfeed for as long as reasonably possible: Human milk contains an ideal balance of nutrients to promote optimal growth and healthy development. Breastfeeding conveys significant short- and long-term health benefits for the infant, promotes mother–infant bonding and provides economic benefits.[39]

Pre-term infants: Breastfeeding reduces the risk or severity of a number of conditions in infancy and later life, including necrotising enterocolitis in preterm infants. Breastfeeding also has significant cognitive benefits, which appear to be more pronounced in preterm infants.

- In addition to breastfeeding for as long as possible, pre-term breastfed infants require iron supplements from four to eight weeks of age.

- Those born at less than thirty-two weeks' gestation usually require fortification of breastmilk with protein and calories in the pre-term period, in order to promote adequate growth.
- If a mother is unable to provide enough breastmilk, breastmilk bank products are available in Australia for pre-term infants or those with serious medical conditions.[40]

Full-term infants: Exclusive breastfeeding is recommended for around six months (see 'Introducing Solids'). It should be continued until at least twelve months of age and beyond for as long as the mother and child desire.

- It is important to note that feeding with expressed breastmilk is both practical and safe, provided the expressed milk is appropriately stored to prevent the risk of bacterial growth.
- It is recommended that breastfed infants receive vitamin D supplements (10 mcg/day) if their mothers are dark skinned or wear a veil because of the potential for vitamin D deficiency in these women.[41]

Breastfeeding mothers and nutritional intake: Good nutrition is important for the health and wellbeing of all women and particularly for breastfeeding women, who have additional nutritional requirements.

- Breastfeeding mothers have increased energy needs, and typically require an additional 2,000 to 2,100 kJ/day. However, this requirement will vary depending on the mother's level of milk production, rate of postpartum weight loss and physical activity levels.[42] Ensuring adequate energy levels may assist in prolonging breastmilk production, allowing breastfeeding to continue.
- Diets containing less than 20 per cent of energy from fat are not recommended for breastfeeding women, because they may affect the fat content of breastmilk.[43]

- Vegan or vegetarian mothers or mothers who follow other forms of restrictive diets have a greater risk for nutrient deficiencies, including iron, zinc, calcium and vitamin B12, and may need referral to a dietician to maximise the nutritional quality of their breastmilk.[44]
- Iodine supplementation should be continued for the duration of breastfeeding.
- Breastfeeding women should aim for nine servings of breads and cereals (preferably wholegrain) each day.[45]
- Exclusion of allergenic foods from the maternal diet has not been shown to prevent allergies in the offspring, and therefore they do not need to be avoided by the breastfeeding mother.[46]
- Breastfeeding may promote postpartum weight loss. However, many women fail to return to their pre-pregnancy weight within six months, and this has been linked to the development of obesity. Modest and gradual weight loss in women who remain overweight may be beneficial.

Introducing solids: While infant feeding guidelines recommend exclusive breastfeeding until around six months of age, emerging evidence suggests that introducing solids after seventeen weeks and before six months of age (while continuing breastfeeding) has potential benefits in terms of reducing the risk of some food allergies.[47-49]

- The Australasian Society of Clinical Immunology and Allergy (ASCIA) has advised relaxing recommendations to avoid certain food groups and allowing the introduction of solid foods after seventeen weeks, while continuing breastfeeding. It is important that the introduction of solids does not occur before seventeen weeks.
- Allergenic foods do not need to be avoided during the introduction of solid foods, unless the child shows an allergic reaction to it. Research is currently underway investigating

the timing of introduction of allergenic foods, including pea-nuts and egg, into the diet of infants to reduce the risks of food-allergy development.

- The order and timing of first foods do not appear to be important, providing that the first foods are nutrient dense and iron fortified. Iron is particularly important to protect against iron-deficiency anaemia and to promote good cogni-tive development in the infant. Good sources of iron include iron-enriched infant cereals, as well as pureed meat, poultry and fish.

- If infant formula is required in the first months of life before solid foods are introduced, there is evidence that hydrolysed formulas may reduce the risk of allergic disease in high-risk infants (i.e. those with a family history of allergy). [50]

3. INFANT – SIX TO TWELVE MONTHS

Introduce a wide range of solid foods: Introducing a wide range of solid foods from the five food groups, with an emphasis on iron-rich foods, is important to promote growth and develop-ment in the infant. It may also assist the child to choose a broader range of foods in later life.

- Continued breastfeeding is encouraged and should be sup-ported. However, while breastmilk continues to be a major source of nutrients, by this stage it no longer provides all of the required nutrients and energy for growth and develop-ment, and the infant's appetite is unlikely to be satisfied by breastmilk alone.

- Most infants can manage finger foods by eight months of age.

- Most infants are willing to accept new textures and flavours, so it is important to gradually introduce new food tastes and textures (from pureed to lumpy and then to normal textures) during this time.

- Infants should not consume foods with added sugar, salt, honey (which may contain clostridium botulinum) or foods high in saturated fat.
- Cow's milk as a drink should be avoided. Feeding infants with whole cow's milk before twelve months of age is associated with an increased incidence of iron deficiency, which can affect the production of hemoglobin and red blood cells and increase the risk of anaemia. Consumption of cow's milk also reduces the bioavailability of non-haem iron provided by other foods, and may be associated with occult loss of blood from the gastrointestinal tract.[51]

4. TODDLER – ONE TO THREE YEARS

Vary the child's diet and establish positive eating behaviours: A variety of nutritious foods, plus daily physical activity, are important for the ongoing development and growth of the child. Eating behaviours are also formed during this phase of life and should be modelled on a regular routine, eating as part of the family unit, with the size of servings tailored to the child's appetite.

- Solid foods should provide an increasing proportion of nutrients after the age of twelve months. Toddlers typically require small, frequent and nutrient-dense meals, which should be consistent with the Australian and New Zealand dietary guidelines.
- Toddlers may go through a picky eating stage, but it is important to continue to offer a wide variety of foods and regular meals to promote healthy eating habits, behaviours and eating patterns.
- Special milks for toddlers are not required for healthy children. Water and pasteurised full-cream milk are recommended drinks at this time.

- Sugar-sweetened drinks and fruit juice should be avoided or limited/diluted. Coffee, tea and other caffeinated drinks are unsuitable for toddlers.
- Consumption of nutrient-poor foods with high levels of saturated fat, sugar, and/or salt (e.g. potato chips, cakes, biscuits and confectionery) should be avoided or limited.
- Parents should also follow healthy eating and physical activity guidelines, as children's eating and lifestyle behaviours are strongly influenced by parenting practices during this period of life.[52]

CONCLUSION

Nutrition and lifestyle factors throughout pre-conception, pregnancy, infancy and early childhood have a profound influence on a child's development and long-term health. The researchers who put together these recommendations add:

> 'The practical, evidence-based recommendations contained in this report are designed to assist parents and healthcare professionals in their efforts to maximise this critical window of opportunity when the foundations of future health are created'.[53]

As the findings of further research become available, additional early-life nutrition recommendations should be formulated and made widely available as part of the preventative health policy agenda in both Australia and New Zealand.

Notes

Chapter 1: A Pilgrimage

1. Barker DJ, Osmond C. Infant mortality, childhood nutrition, and ischaemic heart disease in England and Wales. Lancet 1986; 1:1077–81.
2. Barker D, Winter P, Osmond C, Margetts B, Simmonds S. Weight in infancy and death from ischemic heart disease. Lancet 1989; ii:577.
3. Forsdahl A. Are poor living conditions in childhood and adolescence an important risk factor for arteriosclerotic heart disease? Br J. Prev Soc Med 1977; 31:91–5.
4. Prescott S, Macaubas C, Smallacombe T, Holt B, Sly P, Loh R, et al. Development of allergen-specific T-cell memory in atopic and normal children. Lancet 1999; 353(9148):196–200.

Chapter 2: Striving for a common global vision

1. NCD Facts. The NCD Alliance. [cited October 2014] <http://www.ncdalliance.org/ncd-facts>.
2. Political Declaration of the High-level Meeting of the General Assembly on the Prevention and Control of Non-communicable Diseases. United Nations General Assembly, New York, September 2011. [cited October 2013] <http://www.un.org/en/ga/ncdmeeting2011/>.
3. ibid.
4. Bloom DE, Cafiero E, Jané-Llopis E, Abrahams-Gessel S, Bloom LR, Fathima S, et al. The Global Economic Burden of Non-communicable Diseases, World Economic Forum, Geneva, 2011, [cited October 2014] <www.weforum.org/EconomicsOfNCD>.
5. Political Declaration on the Prevention and Control of Non-communicable Diseases, United Nations General Assembly.
6. Prescott SL. The Allergy Epidemic: A Mystery of Modern Life. Perth: UWA Publishing; 2011.

7. Hanson M, Gluckman P, Nutbeam D, Hearn J. Priority actions for the non-communicable disease crisis. Lancet 2011; 378:566-7.
8. World Allergy Organization (WAO) White Book on Allergy. Eds: Ruby Pawankar, Giorgio Walter Canonica, Stephen T. Holgate and Richard F. Lockey. Milwaukee, Wisconsin: World Allergy Organization <http://www.worldallergy.org>, 2011.
9. Political Declaration on the Prevention and Control of Non-communicable Diseases, United Nations General Assembly.
10. UN Food Security. President Obama, Secretary Clinton, and Secretary Sebelius discuss global health issues during UNGA [cited October 2013] <http://www.un-foodsecurity.org/node/1213>.
11. Political Declaration on the Prevention and Control of Non-communicable Diseases, United Nations General Assembly.
12. M. Chan, World Health Organization Noncommunicable diseases damage health, including economic health [cited October 2013] <http://www.who.int/dg/speeches/2011/un_ncds_09_19/en/>.
13. Political Declaration on the Prevention and Control of Non-communicable Diseases, United Nations General Assembly.

Chapter 3: Early Life: A critical time of risk and a critical time of opportunity

1. Gluckman P, Hanson M. Fat, Fate, and Disease. Oxford: Oxford University Press; 2012.
2. Blencowe H, Cousens S, Oestergaard MZ, Chou D, Moller AB, Narwal R, et al. National, regional, and worldwide estimates of preterm birth rates in the year 2010 with time trends since 1990 for selected countries: a systematic analysis and implications. Lancet 2012; 379:2162-72.
3. Hanson M, Bower C, Milne E, de Klerk N, Kurinczuk JJ. Assisted reproductive technologies and the risk of birth defects—a systematic review. Hum Reprod 2005; 20:328-38.
4. Central Intelligence Agency (CIA) World Fact Book, [cited 28 October 2012] <https://www.cia.gov/library/publications/the-world-factbook/rankorder/2102rank.html>.
5. Volk T, Atkinson J. Is child death the crucible of human evolution? Journal of Social, Evolutionary, and Cultural Psychology 2008; Proceedings of the 2nd Annual Meeting of the NorthEastern Evolutionary Psychology Society:247-60.
6. Marshall BJ, Warren JR. Unidentified curved bacilli on gastric epithelium in active chronic gastritis. Lancet 1983; 1:1273-5.
7. Marshall BJ, Warren JR. Unidentified curved bacilli in the stomach of patients with gastritis and peptic ulceration. Lancet 1984; 1:1311-5.

8. Oransky I. Failure is an option. Cosmos 2012; 47:98.

Chapter 4: The early-life origins of weight gain and obesity

1. Ogden CL, Carroll MD, Kit BK, Flegal KM. Prevalence of obesity in the United States, 2009-2010. NCHS Data Brief 2012:1-8.
2. Australian Bureau of Statistics. Australian Health Survey 2011-12; Released 29/10/2012.
3. Gluckman P, Hanson M. Fat, Fate, and Disease. Oxford: Oxford University Press; 2012.
4. Hoek HW. Incidence, prevalence and mortality of anorexia nervosa and other eating disorders. Curr Opin Psychiatry 2006; 19:389-94.
5. Caulfield T. The Cure for Everything: Untangling Twisted Messages About Health, Fitness and Happiness. Boston, Massachussetts: Beacon Press; 2012.
6. Tardif SD, Power ML, Ross CN, Rutherford JN, Layne-Colon DG, Paulik MA. Characterization of obese phenotypes in a small nonhuman primate, the common marmoset (Callithrix jacchus). Obesity (Silver Spring) 2009; 17:1499-505.
7. Power ML, Ross CN, Schulkin J, Tardif SD. The Development of Obesity Begins at an Early Age in Captive Common Marmosets (Callithrix jacchus). American Journal of Primatology 2012; 74:261-9.
8. Selhub EM, Logan AC. Your Brain on Nature: The Science of Nature's Influence on Your Health, Happiness and Vitality. Mississauga, Canada: Wiley; 2012.
9. Altenburg TM, Chinapaw MJ, van der Knaap ET, Brug J, Manios Y, Singh AS. Longer sleep–slimmer kids: the ENERGY-project. PLoS One 2013; 8:e59522.
10. Keith SW, Redden DT, Katzmarzyk PT, Boggiano MM, Hanlon EC, Benca RM, et al. Putative contributors to the secular increase in obesity: exploring the roads less traveled. Int J Obes (Lond) 2006; 30:1585-94.
11. Archer T, Fredriksson A, Schtutz E, Kostrzewa RM. Influence of Physical Exercise on Neuroimmunological Functioning and Health: Aging and Stress. Neurotox Res 2011; 20:69–83.
12. Petersen AM, Pedersen BK. The anti-inflammatory effect of exercise. J Appl Physiol 2005; 98:1154-62.
13. McFarlin BK, Flynn MG, Campbell WW, Craig BA, Robinson JP, Stewart LK, et al. Physical activity status, but not age, influences inflammatory biomarkers and toll-like receptor 4. J Gerontol A Biol Sci Med Sci 2006; 61:388-93.
14. Petersen AM, Pedersen BK. The anti-inflammatory effect of exercise. J Appl Physiol 2005; 98:1154-62.

15. ibid.
16. Turnbaugh PJ, Ley RE, Mahowald MA, Magrini V, Mardis ER, Gordon JI. An obesity-associated gut microbiome with increased capacity for energy harvest. Nature 2006; 444:1027-31.
17. Bluher S, Mantzoros CS. Leptin in humans: lessons from translational research. Am J Clin Nutr 2009; 89:991S-7S.
18. Fontana L, Meyer TE, Klein S, Holloszy JO. Long-term low-calorie low-protein vegan diet and endurance exercise are associated with low cardiometabolic risk. Rejuvenation Res 2007; 10:225-34.
19. Fontana L, Klein S. Aging, adiposity, and calorie restriction. JAMA 2007; 297:986-94.
20. Vanlint S. Vitamin D and obesity. Nutrients 2013; 5:949-56.
21. Belenchia AM, Tosh AK, Hillman LS, Peterson CA. Correcting vitamin D insufficiency improves insulin sensitivity in obese adolescents: a randomized controlled trial. Am J Clin Nutr 2013; 97:774-81.
22. Loos RJ, Bouchard C. FTO: the first gene contributing to common forms of human obesity. Obes Rev 2008; 9:246-50.
23. Morgan HD, Sutherland HG, Martin DI, Whitelaw E. Epigenetic inheritance at the agouti locus in the mouse. Nat Genet 1999; 23:314-8.
24. Waterland RA, Jirtle RL. Transposable elements: targets for early nutritional effects on epigenetic gene regulation. Mol Cell Biol 2003; 23:5293-300.
25. Waterland RA, Travisano M, Tahiliani KG. Diet-induced hypermethylation at agouti viable yellow is not inherited transgenerationally through the female. FASEB J 2007; 21:3380-5.
26. Dolinoy DC, Huang D, Jirtle RL. Maternal nutrient supplementation counteracts bisphenol A-induced DNA hypomethylation in early development. Proc Natl Acad Sci USA 2007; 104:13056-61.
27. Chevrier J, Gunier RB, Bradman A, Holland NT, Calafat AM, Eskenazi B, et al. Maternal urinary bisphenol A during pregnancy and maternal and neonatal thyroid function in the CHAMACOS study. Environ Health Perspect 2013; 121(1): 138-144.
28. Braun JM, Kalkbrenner AE, Calafat AM, Yolton K, Ye X, Dietrich KN, et al. Impact of early-life bisphenol A exposure on behavior and executive function in children. Pediatrics 2011; 128:873-82.
29. Trasande L, Attina TM, Blustein J. Association between urinary bisphenol A concentration and obesity prevalence in children and adolescents. JAMA 2012; 308:1113-21.

30. Wang HX, Zhou Y, Tang CX, Wu JG, Chen Y, Jiang QW. Association between bisphenol A exposure and body mass index in Chinese school children: a cross-sectional study. Environ Health 2012; 11:79.
31. Ravelli GP, Stein ZA, Susser MW. Obesity in young men after famine exposure in utero and early infancy. N Engl J Med 1976; 295:349-53.
32. Vickers MH, Breier BH, Cutfield WS, Hofman PL, Gluckman PD. Fetal origins of hyperphagia, obesity, and hypertension and postnatal amplification by hypercaloric nutrition. Am J Physiol Endocrinol Metab 2000; 279:E83-7.
33. Brooks AA, Johnson MR, Steer PJ, Pawson ME, Abdalla HI. Birth weight: nature or nurture? Early Hum Dev 1995; 42:29-35.
34. Roseboom T, de Rooij S, Painter R. The Dutch famine and its long-term consequences for adult health. Early Hum Dev 2006; 82:485-91.
35. Roseboom TJ, van der Meulen JH, Osmond C, Barker DJ, Ravelli AC, Bleker OP. Plasma lipid profiles in adults after prenatal exposure to the Dutch famine. Am J Clin Nutr 2000; 72:1101-6.
36. Ravelli AC, van der Meulen JH, Michels RP, Osmond C, Barker DJ, Hales CN, et al. Glucose tolerance in adults after prenatal exposure to famine. Lancet 1998; 351:173-7.
37. Roseboom T, de Rooij S, Painter R. The Dutch famine. Early Hum Dev 2006; 82:485-91.
38. Roseboom TJ, van der Meulen JH, Osmond C, Barker DJ, Ravelli AC, Schroeder-Tanka JM, et al. Coronary heart disease after prenatal exposure to the Dutch famine, 1944-45. Heart 2000; 84:595-8.
39. Hoek HW, Brown AS, Susser E. The Dutch famine and schizophrenia spectrum disorders. Soc Psychiatry Psychiatr Epidemiol 1998; 33:373-9.
40. Roseboom T, de Rooij S, Painter R. The Dutch famine. Early Hum Dev 2006; 82:485-91.
41. Heijmans BT, Tobi EW, Stein AD, Putter H, Blauw GJ, Susser ES, et al. Persistent epigenetic differences associated with prenatal exposure to famine in humans. Proc Natl Acad Sci USA 2008; 105:17046-9.
42. ibid.
43. Gillman MW, Rifas-Shiman S, Berkey CS, Field AE, Colditz GA. Maternal gestational diabetes, birth weight, and adolescent obesity. Pediatrics 2003; 111:e221-6.
44. ibid.
45. Kral JG, Biron S, Simard S, Hould FS, Lebel S, Marceau S, et al. Large maternal weight loss from obesity surgery prevents transmission

of obesity to children who were followed for 2 to 18 years. Pediatrics 2006; 118:e1644-9.

46. Patel MS, Srinivasan M. Metabolic programming in the immediate postnatal life. Ann Nutr Metab 2011; 58 Suppl 2:18-28.

47. Patel MS, Srinivasan M, Laychock SG. Metabolic programming: Role of nutrition in the immediate postnatal life. J Inherit Metab Dis 2009; 32:218-28.

48. Fall CH, Borja JB, Osmond C, Richter L, Bhargava SK, Martorell R, et al. Infant-feeding patterns and cardiovascular risk factors in young adulthood: data from five cohorts in low- and middle-income countries. Int J Epidemiol 2011; 40:47-62.

49. Singhal A, Cole TJ, Fewtrell M, Lucas A. Breastmilk feeding and lipoprotein profile in adolescents born preterm: follow-up of a prospective randomised study. Lancet 2004; 363:1571-8.

50. Jonsdottir OH, Thorsdottir I, Hibberd PL, Fewtrell MS, Wells JC, Palsson GI, et al. Timing of the introduction of complementary foods in infancy: a randomized controlled trial. Pediatrics 2012; 130:1038-45.

Chapter 5: The early-life origins of diabetes and metabolic liver disease

1. Harris MI, Flegal KM, Cowie CC, Eberhardt MS, Goldstein DE, Little RR, et al. Prevalence of diabetes, impaired fasting glucose, and impaired glucose tolerance in U.S. adults. The Third National Health and Nutrition Examination Survey, 1988-1994. Diabetes Care 1998; 21:518-24.

2. World Health Organization: 10 Facts about Diabetes [cited October 2014] < http://www.who.int/features/factfiles/diabetes/facts/en/>.

3. Yan S, Li J, Li S, Zhang B, Du S, Gordon-Larsen P, et al. The expanding burden of cardiometabolic risk in China: the China Health and Nutrition Survey. Obes Rev 2012; 13:810-21.

4. University of North Carolina at Chapel Hill. Child diabetes levels almost four times higher in China than in US. ScienceDaily; 5 July 2012 [cited October 2013] <www.sciencedaily.com/releases/2012/07/120705194138.htm>.

5. McDermott RA, Li M, Campbell SK. Incidence of type 2 diabetes in two Indigenous Australian populations: a 6-year follow-up study. Med J Aust 2010; 192:562-5.

6. Karter AJ, Schillinger D, Adams AS, Moffet HH, Liu J, Adler NE, et al. Elevated rates of diabetes in Pacific Islanders and Asian subgroups: The Diabetes Study of Northern California (DISTANCE). Diabetes Care 2013; 36:574-9.

7.   Lawrence JM, Contreras R, Chen W, Sacks DA. Trends in the prevalence of preexisting diabetes and gestational diabetes mellitus among a racially/ethnically diverse population of pregnant women, 1999-2005. Diabetes Care 2008; 31:899-904.

8.   Prescott SL. The Allergy Epidemic: A Mystery of Modern Life. Perth: UWA Publishing; 2011.

9.   Bach JF. The effect of infections on susceptibility to autoimmune and allergic diseases. N Engl J Med 2002; 347:911-20.

10. Gale EA. The rise of childhood type 1 diabetes in the 20th century. Diabetes 2002; 51:3353-61.

11. ibid.

12. ibid.

13. Abela AG, Fava S. Association of incidence of type 1 diabetes with mortality from infectious disease and with antibiotic susceptibility at a country level. Acta Diabetol 2013; 50(6):859-65.

14. Vaarala O. Is the origin of type 1 diabetes in the gut? Immunol Cell Biol 2012; 90:271-6.

15. de Goffau MC, Luopajarvi K, Knip M, Ilonen J, Ruohtula T, Harkonen T, et al. Fecal microbiota composition differs between children with beta-cell autoimmunity and those without. Diabetes 2013; 62:1238-44.

16. Sudo N, Sawamura S, Tanaka K, Aiba Y, Kubo C, Koga Y. The requirement of intestinal bacterial flora for the development of an IgE production system fully susceptible to oral tolerance induction. J Immunol 1997; 159:1739-45.

17. Petrovsky N. Immunomodulation with microbial vaccines to prevent type 1 diabetes mellitus. Nat Rev Endocrinol 2010; 6:131-8.

18. Wen L, Ley RE, Volchkov PY, Stranges PB, Avanesyan L, Stonebraker AC, et al. Innate immunity and intestinal microbiota in the development of Type 1 diabetes. Nature 2008; 455:1109-13.

19. Norris JM. Infant and childhood diet and type 1 diabetes risk: recent advances and prospects. Curr Diab Rep 2010; 10:345-9.

20. University of South Florida. The TEDDY Study. The environmental determinants of diabetes in the young. [cited October 2014] <https://teddy.epi.usf.edu/>.

21. Knip M, Virtanen SM, Becker D, Dupre J, Krischer JP, Akerblom HK. Early feeding and risk of type 1 diabetes: experiences from the trial to reduce insulin-dependent diabetes mellitus in the Genetically at Risk (TRIGR). Am J Clin Nutr 2011; 94:1814S-20S.

22. ibid.

23. National Institute of Diabetes and Digestive and Kidney Diseases, National Institutes of Health (NIH), USA. National Diabetes

Statistics, 2011; <http://diabetes.niddk.nih.gov/dm/pubs/statistics/dm_statistics.pdf>.

24. Grady D. Obesity-linked diabetes in children resists treatment. The New York Times, April 29 2012.

25. Qatanani M, Lazar MA. Mechanisms of obesity-associated insulin resistance: many choices on the menu. Genes Dev 2007; 21:1443-55.

26. Prentki M, Nolan CJ. Islet beta cell failure in type 2 diabetes. J Clin Invest 2006; 116:1802-12.

27. Barker DJ, Hales CN, Fall CH, Osmond C, Phipps K, Clark PM. Type 2 (non-insulin-dependent) diabetes mellitus, hypertension and hyperlipidaemia (syndrome X): relation to reduced fetal growth. Diabetologia 1993; 36:62-7.

28. Hales CN, Barker DJ, Clark PM, Cox LJ, Fall C, Osmond C, et al. Fetal and infant growth and impaired glucose tolerance at age 64. BMJ 1991; 303:1019-22.

29. Whincup PH, Kaye SJ, Owen CG, Huxley R, Cook DG, Anazawa S, et al. Birth weight and risk of type 2 diabetes: a systematic review. JAMA 2008; 300:2886-97.

30. Economides DL, Proudler A, Nicolaides KH. Plasma insulin in appropriate- and small-for-gestational-age fetuses. Am J Obstet Gynecol 1989; 160:1091-4.

31. Brufani C, Grossi A, Fintini D, Tozzi A, Nocerino V, Patera PI, et al. Obese children with low birth weight demonstrate impaired beta-cell function during oral glucose tolerance test. J Clin Endocrinol Metab 2009; 94:4448-52.

32. Van Assche FA, De Prins F, Aerts L, Verjans M. The endocrine pancreas in small-for-dates infants. Br J Obstet Gynaecol 1977; 84:751-3.

33. Ng SF, Lin RC, Laybutt DR, Barres R, Owens JA, Morris MJ. Chronic high-fat diet in fathers programs beta-cell dysfunction in female rat offspring. Nature 2010; 467:963-6.

34. ibid.

35. Patel MS, Srinivasan M. Metabolic programming: causes and consequences. J Biol Chem 2002; 277:1629-32.

36. Ferrara A. Increasing prevalence of gestational diabetes mellitus: a public health perspective. Diabetes Care 2007; 30 Suppl 2:S141-6.

37. Singh R, Pearson E, Avery PJ, McCarthy MI, Levy JC, Hitman GA, et al. Reduced beta cell function in offspring of mothers with young-onset type 2 diabetes. Diabetologia 2006; 49:1876-80.

38. Portha B, Chavey A, Movassat J. Early-life origins of type 2 diabetes: fetal programming of the beta-cell mass. Exp Diabetes Res 2011; 2011:105076.

39. Sepp E, Julge K, Vasar M, Naaber P, Bjorksten B, Mikelsaar M. Intestinal microflora of Estonian and Swedish infants. Acta Paediatr 1997; 86:956-61.

40. Larsen N, Vogensen FK, van den Berg FW, Nielsen DS, Andreasen AS, Pedersen BK, et al. Gut microbiota in human adults with type 2 diabetes differs from non-diabetic adults. PLoS One 2010; 5:e9085.

41. Luoto R, Laitinen K, Nermes M, Isolauri E. Impact of maternal probiotic-supplemented dietary counselling on pregnancy outcome and prenatal and postnatal growth: a double-blind, placebo-controlled study. Br J Nutr 2010; 103:1792-9.

42. Luoto R, Laitinen K, Nermes M, Isolauri E. Impact of maternal probiotic-supplemented dietary counseling during pregnancy on colostrum adiponectin concentration: a prospective, randomized, placebo-controlled study. Early Hum Dev 2012; 88:339-44.

43. Aaltonen J, Ojala T, Laitinen K, Poussa T, Ozanne S, Isolauri E. Impact of maternal diet during pregnancy and breastfeeding on infant metabolic programming: a prospective randomized controlled study. Eur J Clin Nutr 2011; 65:10-9.

44. ibid.

45. Tilg H, Kaser A. Gut microbiome, obesity, and metabolic dysfunction. J Clin Invest 2011; 121:2126-32.

46. Prescott SL. Early-life environmental determinants of allergic diseases and the wider pandemic of inflammatory noncommunicable diseases. J Allergy Clin Immunol 2013; 131:23-30.

47. Sen S, Simmons RA. Maternal antioxidant supplementation prevents adiposity in the offspring of Western diet-fed rats. Diabetes 2010; 59:3058-65.

48. Chen JH, Hales CN, Ozanne SE. DNA damage, cellular senescence and organismal ageing: causal or correlative? Nucleic Acids Res 2007; 35:7417-28.

49. Jennings BJ, Ozanne SE, Dorling MW, Hales CN. Early growth determines longevity in male rats and may be related to telomere shortening in the kidney. FEBS Lett 1999; 448:4-8.

50. Tarry-Adkins JL, Chen JH, Smith NS, Jones RH, Cherif H, Ozanne SE. Poor maternal nutrition followed by accelerated postnatal growth leads to telomere shortening and increased markers of cell senescence in rat islets. FASEB J 2009; 23:1521-8.

51. Vajro P, Lenta S, Socha P, Dhawan A, McKiernan P, Baumann U, et al. Diagnosis of nonalcoholic fatty liver disease in children and adolescents: position paper of the ESPGHAN Hepatology Committee. J Pediatr Gastroenterol Nutr 2012; 54:700-13.

52. Hilden M, Christoffersen P, Juhl E, Dalgaard JB. Liver histology in a 'normal' population—examinations of 503 consecutive fatal traffic casualties. Scand J Gastroenterol 1977; 12:593-7.

53. Adler M, Schaffner F. Fatty liver hepatitis and cirrhosis in obese patients. Am J Med 1979; 67:811-6.

54. Ludwig J, Viggiano TR, McGill DB, Oh BJ. Nonalcoholic steatohepatitis: Mayo Clinic experiences with a hitherto unnamed disease. Mayo Clin Proc 1980; 55:434-8.

55. Moran JR, Ghishan FK, Halter SA, Greene HL. Steatohepatitis in obese children: a cause of chronic liver dysfunction. Am J Gastroenterol 1983; 78:374-7.

56. ibid.

57. Schwimmer JB, Deutsch R, Kahen T, Lavine JE, Stanley C, Behling C. Prevalence of fatty liver in children and adolescents. Pediatrics 2006; 118:1388-93.

58. Alavian SM, Mohammad-Alizadeh AH, Esna-Ashari F, Ardalan G, Hajarizadeh B. Non-alcoholic fatty liver disease prevalence among school-aged children and adolescents in Iran and its association with biochemical and anthropometric measures. Liver Int 2009; 29:159-63.

59. McCullough AJ. The clinical features, diagnosis and natural history of nonalcoholic fatty liver disease. Clin Liver Dis 2004; 8:521-33, viii.

60. Alisi A, Panera N, Agostoni C, Nobili V. Intrauterine growth retardation and nonalcoholic fatty liver disease in children. Int J Endocrinol 2011; 2011:269853.

61. Nobili V, Alisi A, Panera N, Agostoni C. Low birth weight and catch-up-growth associated with metabolic syndrome: a ten year systematic review. Pediatr Endocrinol Rev 2008; 6:241-7.

62. Nobili V, Bedogni G, Alisi A, Pietrobattista A, Alterio A, Tiribelli C, et al. A protective effect of breastfeeding on the progression of non-alcoholic fatty liver disease. Arch Dis Child 2009; 94:801-5.

63. Bray GA, Popkin BM. Calorie-sweetened beverages and fructose: what have we learned 10 years later. Pediatr Obes 2013; 8:242-8.

64. Johnson RK, Appel LJ, Brands M, Howard BV, Lefevre M, Lustig RH, et al. Dietary sugars intake and cardiovascular health: a scientific statement from the American Heart Association. Circulation 2009; 120:1011-20.

65. McCann MF, Baydar N, Williams RL. Consumption of soft drinks and other sweet drinks by WIC infants. Am J Public Health 2008; 98:1735.

66. Vartanian LR, Schwartz MB, Brownell KD. Effects of soft drink consumption on nutrition and health: a systematic review and meta-analysis. Am J Public Health 2007; 97:667-75.

67. Kranz S, Smiciklas-Wright H, Siega-Riz AM, Mitchell D. Adverse effect of high added sugar consumption on dietary intake in American preschoolers. J Pediatr 2005; 146:105-11.

68. Le KA, Tappy L. Metabolic effects of fructose. Curr Opin Clin Nutr Metab Care 2006; 9:469-75.

69. Havel PJ. Dietary fructose: implications for dysregulation of energy homeostasis and lipid/carbohydrate metabolism. Nutr Rev 2005; 63:133-57.

70. Gross LS, Li L, Ford ES, Liu S. Increased consumption of refined carbohydrates and the epidemic of type 2 diabetes in the United States: an ecologic assessment. Am J Clin Nutr 2004; 79:774-9.

71. Payne AN, Chassard C, Lacroix C. Gut microbial adaptation to dietary consumption of fructose, artificial sweeteners and sugar alcohols: implications for host-microbe interactions contributing to obesity. Obes Rev 2012; 13:799-809.

72. Kaumi T, Hirano T, Odaka H, Ebara T, Amano N, Hozumi T, et al. VLDL triglyceride kinetics in Wistar fatty rats, an animal model of NIDDM: effects of dietary fructose alone or in combination with pioglitazone. Diabetes 1996; 45:806-11.

73. Stanhope KL, Schwarz JM, Keim NL, Griffen SC, Bremer AA, Graham JL, et al. Consuming fructose-sweetened, not glucose-sweetened, beverages increases visceral adiposity and lipids and decreases insulin sensitivity in overweight/obese humans. J Clin Invest 2009; 119:1322-34.

74. Duffey KJ, Popkin BM. High-fructose corn syrup: is this what's for dinner? Am J Clin Nutr 2008; 88:1722S-32S.

75. Singhal A, Fewtrell M, Cole TJ, Lucas A. Low nutrient intake and early growth for later insulin resistance in adolescents born preterm. Lancet 2003; 361:1089-97.

76. Nobili V, Alisi A, Della Corte C, Rise P, Galli C, Agostoni C, et al. Docosahexaenoic acid for the treatment of fatty liver: Randomised controlled trial in children. Nutr Metab Cardiovasc Dis 2012.

77. Cankurtaran M, Kav T, Yavuz B, Shorbagi A, Halil M, Coskun T, et al. Serum vitamin-E levels and its relation to clinical features in nonalcoholic fatty liver disease with elevated ALT levels. Acta Gastroenterol Belg 2006; 69:5-11.

Chapter 6: The early-life origins of heart disease and cardiovascular disorders

1. Global status report on noncommunicable disaeses 2010. Geneva, World Health Organization, 2011. < http://www.who.int/nmh/ publications/ncd_report_full_en.pdf>.

2. Ozcetin M, Celikyay ZR, Celik A, Yilmaz R, Yerli Y, Erkorkmaz U. The importance of carotid artery stiffness and increased intima-media thickness in obese children. S Afr Med J 2012; 102:295-9.

3. Iannuzzi A, Licenziati MR, Acampora C, Salvatore V, Auriemma L, Romano ML, et al. Increased carotid intima-media thickness and stiffness in obese children. Diabetes Care 2004; 27:2506-8.

4. McCrindle BW, Urbina EM, Dennison BA, Jacobson MS, Steinberger J, Rocchini AP, et al. Drug therapy of high-risk lipid abnormalities in children and adolescents: a scientific statement from the American Heart Association Atherosclerosis, Hypertension, and Obesity in Youth Committee, Council of Cardiovascular Disease in the Young, with the Council on Cardiovascular Nursing. Circulation 2007; 115:1948-67.

5. Huang RC, Mori TA, Burke V, Newnham J, Stanley FJ, Landau LI, et al. Synergy between adiposity, insulin resistance, metabolic risk factors, and inflammation in adolescents. Diabetes Care 2009; 32:695-701.

6. Forsdahl A. Are poor living conditions in childhood and adolescence an important risk factor for arteriosclerotic heart disease? Br J. Prev Soc Med 1977; 31:91–5.

7. Chobanian AV, Bakris GL, Black HR, Cushman WC, Green LA, Izzo JL, Jr., et al. Seventh report of the Joint National Committee on Prevention, Detection, Evaluation, and Treatment of High Blood Pressure. Hypertension 2003; 42:1206-52.

8. Kearney PM, Whelton M, Reynolds K, Muntner P, Whelton PK, He J. Global burden of hypertension: analysis of worldwide data. Lancet 2005; 365:217-23.

9. Heidenreich PA, Trogdon JG, Khavjou OA, Butler J, Dracup K, Ezekowitz MD, et al. Forecasting the future of cardiovascular disease in the United States: a policy statement from the American Heart Association. Circulation 2011; 123:933-44.

10. Yoon SS, Ostchega Y, Louis T. Recent trends in the prevalence of high blood pressure and its treatment and control, 1999-2008. NCHS Data Brief 2010:1-8.

11. Lloyd-Jones D, Adams RJ, Brown TM, Carnethon M, Dai S, De Simone G, et al. Heart disease and stroke statistics–2010 update: a report from the American Heart Association. Circulation 2010; 121:e46-e215.

12. Chockalingam A. Impact of World Hypertension Day. Can J Cardiol 2007; 23:517-9.

13. Zhao Y, Yan H, Marshall RJ, Dang S, Yang R, Li Q, et al. Trends in population blood pressure and prevalence, awareness, treatment,

and control of hypertension among middle-aged and older adults in a rural area of Northwest China from 1982 to 2010. PLoS One 2013; 8:e61779.

14. Lloyd-Jones D, Adams RJ, Brown TM, Carnethon M, Dai S, De Simone G, et al. Heart disease and stroke statistics–2010 update: a report from the American Heart Association. Circulation 2010; 121:e46-e215.

15. Heidenreich PA, Trogdon JG, Khavjou OA, Butler J, Dracup K, Ezekowitz MD, et al. Forecasting the future of cardiovascular disease in the United States: a policy statement from the American Heart Association. Circulation 2011; 123:933-44.

16. ibid.

17. Nichols M, Townsend N, Luengo-Fernandez R, Leal J, Gray A, Scarborough P, et al. European Cardiovascular Disease Statistics 2012. European Heart Network, Brussels, European Society of Cardiology, Sophia Antipolis. 2012.

18. Lifton RP, Gharavi AG, Geller DS. Molecular mechanisms of human hypertension. Cell 2001; 104:545-56.

19. Ehret GB, Munroe PB, Rice KM, Bochud M, Johnson AD, Chasman DI, et al. Genetic variants in novel pathways influence blood pressure and cardiovascular disease risk. Nature 2011; 478:103-9.

20. Navar LG. Counterpoint: Activation of the intrarenal renin-angiotensin system is the dominant contributor to systemic hypertension. J Appl Physiol 2010; 109:1998-2000; discussion 15.

21. Esler M, Lambert E, Schlaich M. Point: Chronic activation of the sympathetic nervous system is the dominant contributor to systemic hypertension. J Appl Physiol 2010; 109:1996-8; discussion 2016.

22. Savoia C, Schiffrin EL. Inflammation in hypertension. Curr Opin Nephrol Hypertens 2006; 15:152-8.

23. Crowley SD. The Cooperative Roles of Inflammation and Oxidative Stress in the Pathogenesis of Hypertension. Antioxid Redox Signal 2013.

24. Mesas AE, Leon-Munoz LM, Rodriguez-Artalejo F, Lopez-Garcia E. The effect of coffee on blood pressure and cardiovascular disease in hypertensive individuals: a systematic review and meta-analysis. Am J Clin Nutr 2011; 94:1113-26.

25. Vaidya A, Forman JP. Vitamin D and hypertension: current evidence and future directions. Hypertension 2010; 56:774-9.

26. Hart PH, Gorman S, Finlay-Jones JJ. Modulation of the immune system by UV radiation: more than just the effects of vitamin D? Nat Rev Immunol 2011; 11:584-96.

27. Ospina MB, Bond K, Karkhaneh M, Tjosvold L, Vandermeer B, Liang Y, et al. Meditation practices for health: state of the research. Evid Rep Technol Assess (Full Rep) 2007:1-263.

28. Fogari R, Zoppi A, Corradi L, Preti P, Mugellini A, Lazzari P, et al. Effect of body weight loss and normalization on blood pressure in overweight non-obese patients with stage 1 hypertension. Hypertens Res 2010; 33:236-42.

29. Haslam DW, James WP. Obesity. Lancet 2005; 366:1197-209.

30. He FJ, Li J, Macgregor GA. Effect of longer term modest salt reduction on blood pressure: Cochrane systematic review and meta-analysis of randomised trials. BMJ 2013; 346:f1325.

31. Dickinson HO, Mason JM, Nicolson DJ, Campbell F, Beyer FR, Cook JV, et al. Lifestyle interventions to reduce raised blood pressure: a systematic review of randomized controlled trials. J Hypertens 2006; 24:215-33.

32. Whelton PK, He J, Appel LJ, Cutler JA, Havas S, Kotchen TA, et al. Primary prevention of hypertension: clinical and public health advisory from the National High Blood Pressure Education Program. JAMA 2002; 288:1882-8.

33. Appel LJ, Brands MW, Daniels SR, Karanja N, Elmer PJ, Sacks FM. Dietary approaches to prevent and treat hypertension: a scientific statement from the American Heart Association. Hypertension 2006; 47:296-308.

34. Aucott L, Rothnie H, McIntyre L, Thapa M, Waweru C, Gray D. Long-term weight loss from lifestyle intervention benefits blood pressure?: a systematic review. Hypertension 2009; 54:756-62.

35. Lawlor DA, Smith GD. Early life determinants of adult blood pressure. Curr Opin Nephrol Hypertens 2005; 14:259-64.

36. Law CM, Shiell AW, Newsome CA, Syddall HE, Shinebourne EA, Fayers PM, et al. Fetal, infant, and childhood growth and adult blood pressure: a longitudinal study from birth to 22 years of age. Circulation 2002; 105:1088-92.

37. Eriksson J, Forsen T, Tuomilehto J, Osmond C, Barker D. Fetal and childhood growth and hypertension in adult life. Hypertension 2000; 36:790-4.

38. Martin RM, Gunnell D, Smith GD. Breastfeeding in infancy and blood pressure in later life: systematic review and meta-analysis. Am J Epidemiol 2005; 161:15-26.

39. Lawlor DA, Najman JM, Sterne J, Williams GM, Ebrahim S, Davey Smith G. Associations of parental, birth, and early life characteristics with systolic blood pressure at 5 years of age: findings from the

Mater-University study of pregnancy and its outcomes. Circulation 2004; 110:2417-23.

40. Whincup PH, Cook DG, Shaper AG. Early influences on blood pressure: a study of children aged 5-7 years. BMJ 1989; 299:587-91.

41. Ng SF, Lin RC, Laybutt DR, Barres R, Owens JA, Morris MJ. Chronic high-fat diet in fathers programs beta-cell dysfunction in female rat offspring. Nature 2010; 467:963-6.

42. Blake KV, Gurrin LC, Evans SF, Beilin LJ, Landau LI, Stanley FJ, et al. Maternal cigarette smoking during pregnancy, low birth weight and subsequent blood pressure in early childhood. Early Hum Dev 2000; 57:137-47.

43. ibid.

44. Lawlor DA, Najman JM, Sterne J, Williams GM, Ebrahim S, Davey Smith G. Associations of parental, birth, and early life characteristics with systolic blood pressure. Circulation 2004; 110:2417-23.

45. Barker DJ, Gelow J, Thornburg K, Osmond C, Kajantie E, Eriksson JG. The early origins of chronic heart failure: impaired placental growth and initiation of insulin resistance in childhood. Eur J Heart Fail 2010; 12:819-25.

46. Barker DJ, Osmond C, Golding J, Kuh D, Wadsworth ME. Growth in utero, blood pressure in childhood and adult life, and mortality from cardiovascular disease. BMJ 1989; 298:564-7.

47. Brenner BM, Garcia DL, Anderson S. Glomeruli and blood pressure. Less of one, more the other? Am J Hypertens 1988; 1:335-47.

48. ibid.

49. Manalich R, Reyes L, Herrera M, Melendi C, Fundora I. Relationship between weight at birth and the number and size of renal glomeruli in humans: a histomorphometric study. Kidney Int 2000; 58:770-3.

50. Zandi-Nejad K, Luyckx VA, Brenner BM. Adult hypertension and kidney disease: the role of fetal programming. Hypertension 2006; 47:502-8.

51. Woods LL, Weeks DA, Rasch R. Programming of adult blood pressure by maternal protein restriction: role of nephrogenesis. Kidney Int 2004; 65:1339-48.

52. Riviere G, Michaud A, Breton C, VanCamp G, Laborie C, Enache M, et al. Angiotensin-converting enzyme 2 (ACE2) and ACE activities display tissue-specific sensitivity to undernutrition-programmed hypertension in the adult rat. Hypertension 2005; 46:1169-74.

53. Chou HC, Wang LF, Lu KS, Chen CM. Effects of maternal undernutrition on renal angiotensin II and chymase in hypertensive offspring. Acta Histochem 2008; 110:497-504.

54. Keller G, Zimmer G, Mall G, Ritz E, Amann K. Nephron number in patients with primary hypertension. N Engl J Med 2003; 348:101-8.

55. Hughson MD, Douglas-Denton R, Bertram JF, Hoy WE. Hypertension, glomerular number, and birth weight in African Americans and white subjects in the southeastern United States. Kidney Int 2006; 69:671-8.

56. Stewart T, Jung FF, Manning J, Vehaskari VM. Kidney immune cell infiltration and oxidative stress contribute to prenatally programmed hypertension. Kidney Int 2005; 68:2180-8.

57. Wintour EM, Moritz KM, Johnson K, Ricardo S, Samuel CS, Dodic M. Reduced nephron number in adult sheep, hypertensive as a result of prenatal glucocorticoid treatment. J Physiol 2003; 549:929-35.

58. Li G, Xiao Y, Estrella JL, Ducsay CA, Gilbert RD, Zhang L. Effect of fetal hypoxia on heart susceptibility to ischemia and reperfusion injury in the adult rat. J Soc Gynecol Investig 2003; 10:265-74.

59. Hayes EK, Lechowicz A, Petrik JJ, Storozhuk Y, Paez-Parent S, Dai Q, et al. Adverse fetal and neonatal outcomes associated with a life-long high fat diet: role of altered development of the placental vasculature. PLoS One 2012; 7:e33370.

60. Thompson RC, Allam AH, Lombardi GP, Wann LS, Sutherland ML, Sutherland JD, et al. Atherosclerosis across 4000 years of human history: the Horus study of four ancient populations. Lancet 2013; 381:1211-22.

61. Pepys MB, Hirschfield GM. C-reactive protein: a critical update. J Clin Invest 2003; 111:1805-12.

62. Ridker PM, Danielson E, Fonseca FA, Genest J, Gotto AM, Jr., Kastelein JJ, et al. Rosuvastatin to prevent vascular events in men and women with elevated C-reactive protein. N Engl J Med 2008; 359:2195-207.

63. Ridker PM, Buring JE, Cook NR, Rifai N. C-reactive protein, the metabolic syndrome, and risk of incident cardiovascular events: an 8-year follow-up of 14 719 initially healthy American women. Circulation 2003; 107:391-7.

64. Pradhan AD, Manson JE, Rifai N, Buring JE, Ridker PM. C-reactive protein, interleukin 6, and risk of developing type 2 diabetes mellitus. JAMA 2001; 286:327-34.

65. Jenny NS, Yanez ND, Psaty BM, Kuller LH, Hirsch CH, Tracy RP. Inflammation biomarkers and near-term death in older men. Am J Epidemiol 2007; 165:684-95.

66. Pearson TA, Mensah GA, Alexander RW, Anderson JL, Cannon RO, 3rd, Criqui M, et al. Markers of inflammation and cardiovascular disease: application to clinical and public health practice: A statement for healthcare professionals from the Centers for Disease Control and Prevention and the American Heart Association. Circulation 2003; 107:499-511.

67. Lagrand WK, Visser CA, Hermens WT, Niessen HW, Verheugt FW, Wolbink GJ, et al. C-reactive protein as a cardiovascular risk factor: more than an epiphenomenon? Circulation 1999; 100:96-102.

68. Pasceri V, Willerson JT, Yeh ET. Direct proinflammatory effect of C-reactive protein on human endothelial cells. Circulation 2000; 102:2165-8.

69. Zwaka TP, Hombach V, Torzewski J. C-reactive protein-mediated low density lipoprotein uptake by macrophages: implications for atherosclerosis. Circulation 2001; 103:1194-7.

70. McDade TW, Rutherford JN, Adair L, Kuzawa C. Population differences in associations between C-reactive protein concentration and adiposity: comparison of young adults in the Philippines and the United States. Am J Clin Nutr 2009; 89:1237-45.

71. McDade TW, Tallman PS, Madimenos FC, Liebert MA, Cepon TJ, Sugiyama LS, et al. Analysis of variability of high sensitivity C-reactive protein in lowland Ecuador reveals no evidence of chronic low-grade inflammation. Am J Hum Biol 2012; 24:675-81.

72. Ohsawa M, Okayama A, Nakamura M, Onoda T, Kato K, Itai K, et al. CRP levels are elevated in smokers but unrelated to the number of cigarettes and are decreased by long-term smoking cessation in male smokers. Prev Med 2005; 41:651-6.

73. Dietrich T, Garcia RI, de Pablo P, Schulze PC, Hoffmann K. The effects of cigarette smoking on C-reactive protein concentrations in men and women and its modification by exogenous oral hormones in women. Eur J Cardiovasc Prev Rehabil 2007; 14:694-700.

74. O'Loughlin J, Lambert M, Karp I, McGrath J, Gray-Donald K, Barnett TA, et al. Association between cigarette smoking and C-reactive protein in a representative, population-based sample of adolescents. Nicotine Tob Res 2008; 10:525-32.

75. Julia C, Meunier N, Touvier M, Ahluwalia N, Sapin V, Papet I, et al. Dietary patterns and risk of elevated C-reactive protein concentrations 12 years later. Br J Nutr 2013:1-8.

76. Ciubotaru I, Lee YS, Wander RC. Dietary fish oil decreases C-reactive protein, interleukin-6, and triacylglycerol to HDL-cholesterol ratio in postmenopausal women on HRT. J Nutr Biochem 2003; 14:513-21.

77. Muhammad KI, Morledge T, Sachar R, Zeldin A, Wolski K, Bhatt DL. Treatment with ω-3 fatty acids reduces serum C-reactive protein concentration. Clin Lipidol 2011; 6 (6):723-9.

78. Martinelli N, Girelli D, Malerba G, Guarini P, Illig T, Trabetti E, et al. FADS genotypes and desaturase activity estimated by the ratio of arachidonic acid to linoleic acid are associated with inflammation and coronary artery disease. Am J Clin Nutr 2008; 88:941-9.

79. Marik PE, Varon J. Omega-3 dietary supplements and the risk of cardiovascular events: a systematic review. Clin Cardiol 2009; 32:365-72.

80. Rizos EC, Ntzani EE, Bika E, Kostapanos MS, Elisaf MS. Association between omega-3 fatty acid supplementation and risk of major cardiovascular disease events: a systematic review and meta-analysis. JAMA 2012; 308:1024-33.

81. Kones R. Inflammation, C-reactive protein and cardiometabolic risk: how compelling is the potential therapeutic role of n-3 PUFAs in cardiovascular disease? Clin. Lipidol 2011; 6(6):627-30.

82. Visser M, Bouter LM, McQuillan GM, Wener MH, Harris TB. Elevated C-reactive protein levels in overweight and obese adults. JAMA 1999; 282:2131-5.

83. Ridker PM, Rifai N, Rose L, Buring JE, Cook NR. Comparison of C-reactive protein and low-density lipoprotein cholesterol levels in the prediction of first cardiovascular events. N Engl J Med 2002; 347:1557-65.

84. Ott SJ, El Mokhtari NE, Musfeldt M, Hellmig S, Freitag S, Rehman A, et al. Detection of diverse bacterial signatures in atherosclerotic lesions of patients with coronary heart disease. Circulation 2006; 113:929-37.

85. Koren O, Spor A, Felin J, Fak F, Stombaugh J, Tremaroli V, et al. Human oral, gut, and plaque microbiota in patients with atherosclerosis. Proc Natl Acad Sci U S A 2011; 108 Suppl 1:4592-8.

86. Ott SJ, El Mokhtari NE, Musfeldt M, Hellmig S, Freitag S, Rehman A, et al. Detection of diverse bacterial signatures. Circulation 2006; 113:929-37.

87. Manolakis A, Kapsoritakis AN, Potamianos SP. A review of the postulated mechanisms concerning the association of Helicobacter pylori with ischemic heart disease. Helicobacter 2007; 12:287-97.

88. Rosenfeld ME, Campbell LA. Pathogens and atherosclerosis: update on the potential contribution of multiple infectious organisms to the pathogenesis of atherosclerosis. Thromb Haemost 2011; 106:858-67.

89. Loscalzo J. Lipid metabolism by gut microbes and atherosclerosis. Circ Res 2011; 109:127-9.

90. Wang Z, Klipfell E, Bennett BJ, Koeth R, Levison BS, Dugar B, et al. Gut flora metabolism of phosphatidylcholine promotes cardiovascular disease. Nature 2011; 472:57-63.

91. ibid.

92. ibid.

93. Martin FP, Wang Y, Sprenger N, Yap IK, Lundstedt T, Lek P, et al. Probiotic modulation of symbiotic gut microbial-host metabolic interactions in a humanized microbiome mouse model. Mol Syst Biol 2008; 4:157.

94. Swardfager W, Herrmann N, Cornish S, Mazereeuw G, Marzolini S, Sham L, et al. Exercise intervention and inflammatory markers in coronary artery disease: a meta-analysis. Am Heart J 2012; 163:666-76 e1-3.

95. Meier-Ewert HK, Ridker PM, Rifai N, Regan MM, Price NJ, Dinges DF, et al. Effect of sleep loss on C-reactive protein, an inflammatory marker of cardiovascular risk. J Am Coll Cardiol 2004; 43:678-83.

96. ibid.

97. Schwartz SW, Cornoni-Huntley J, Cole SR, Hays JC, Blazer DG, Schocken DD. Are sleep complaints an independent risk factor for myocardial infarction? Ann Epidemiol 1998; 8:384-92.

98. Kripke DF, Garfinkel L, Wingard DL, Klauber MR, Marler MR. Mortality associated with sleep duration and insomnia. Arch Gen Psychiatry 2002; 59:131-6.

99. Appels A, de Vos Y, van Diest R, Hoppner P, Mulder P, de Groen J. Are sleep complaints predictive of future myocardial infarction? Act Nerv Super (Praha) 1987; 29:147-51.

100. Larkin EK, Rosen CL, Kirchner HL, Storfer-Isser A, Emancipator JL, Johnson NL, et al. Variation of C-reactive protein levels in adolescents: association with sleep-disordered breathing and sleep duration. Circulation 2005; 111:1978-84.

101. Tauman R, O'Brien LM, Gozal D. Hypoxemia and obesity modulate plasma C-reactive protein and interleukin-6 levels in sleep-disordered breathing. Sleep Breath 2007; 11:77-84.

102 McDade TW. Early environments and the ecology of inflammation. Proc Natl Acad Sci USA 2012; 109 Suppl 2:17281-8.

103. ibid.

104. Adair LS, Popkin BM, Akin JS, Guilkey DK, Gultiano S, Borja J, et al. Cohort profile: the Cebu longitudinal health and nutrition survey. Int J Epidemiol 2011; 40:619-25.

105. Moe CL, Sobsey MD, Samsa GP, Mesolo V. Bacterial indicators of risk of diarrhoeal disease from drinking-water in the Philippines. Bull World Health Organ 1991; 69:305-17.

106. VanDerslice J, Popkin B, Briscoe J. Drinking-water quality, sanitation, and breast-feeding: their interactive effects on infant health. Bull World Health Organ 1994; 72:589-601.

107. McDade TW, Rutherford J, Adair L, Kuzawa CW. Early origins of inflammation: microbial exposures in infancy predict lower levels of C-reactive protein in adulthood. Proc Biol Sci 2010; 277:1129-37.

108. McDade TW. Early environments and the ecology of inflammation. Proc Natl Acad Sci USA 2012; 109 Suppl 2:17281-8.

109. Gurven M, Kaplan H, Winking J, Eid Rodriguez D, Vasunilashorn S, Kim JK, et al. Inflammation and infection do not promote arterial aging and cardiovascular disease risk factors among lean horticulturalists. PLoS One 2009; 4:e6590.

110. Prescott SL. The Allergy Epidemic: A Mystery of Modern Life. Perth: UWA Publishing; 2011.

111. Prescott SL. Early-life environmental determinants of allergic diseases and the wider pandemic of inflammatory noncommunicable diseases. J Allergy Clin Immunol 2013; 131:23-30.

112. Prescott SL. Early origins of allergic disease: a review of processes and influences during early immune development. Curr Opin Allergy Clin Immunol 2003; 3:125-32.

113. ibid.

114. Danese A, Pariante CM, Caspi A, Taylor A, Poulton R. Childhood maltreatment predicts adult inflammation in a life-course study. Proc Natl Acad Sci U S A 2007; 104:1319-24.

115. Tzoulaki I, Jarvelin MR, Hartikainen AL, Leinonen M, Pouta A, Paldanius M, et al. Size at birth, weight gain over the life course, and low-grade inflammation in young adulthood: northern Finland 1966 Birth Cohort study. Eur Heart J 2008; 29:1049-56.

116. Sattar N, McConnachie A, O'Reilly D, Upton MN, Greer IA, Davey Smith G, et al. Inverse association between birth weight and C-reactive protein concentrations in the MIDSPAN Family Study. Arterioscler Thromb Vasc Biol 2004; 24:583-7.

117. Taylor SE, Lehman BJ, Kiefe CI, Seeman TE. Relationship of early life stress and psychological functioning to adult C-reactive protein in the coronary artery risk development in young adults study. Biol Psychiatry 2006; 60:819-24.

118. Miller GE, Chen E, Fok AK, Walker H, Lim A, Nicholls EF, et al. Low early-life social class leaves a biological residue manifested by decreased glucocorticoid and increased proinflammatory signaling. Proc Natl Acad Sci U S A 2009; 106:14716-21.

119. Miller GE, Chen E. Harsh family climate in early life presages the emergence of a proinflammatory phenotype in adolescence. Psychol Sci 2010; 21:848-56.

120. Metzger MW, McDade TW. Breastfeeding as obesity prevention in the United States: a sibling difference model. Am J Hum Biol 2010; 22:291-6.

121. A statistical overview of Aboriginal and Torres Strait Islander peoples in Australia: Social Justice Report 2008, Australian Human Rights Commission.

122. ibid.

123. Marmot M. Social determinants of health inequalities. Lancet 2005; 365:1099-104.

124. ibid.

Chapter 7: Brain, behaviour and mood

1. Stiles J, Jernigan TL. The basics of brain development. Neuropsychol Rev 2010; 20:327-48.

2. van Os J, Kapur S. Schizophrenia. Lancet 2009; 374:635-45.

3. Kircher TT, Thienel R. Functional brain imaging of symptoms and cognition in schizophrenia. Prog Brain Res 2005; 150:299-308.

4. Insel TR. Rethinking schizophrenia. Nature 2010; 468:187-93.

5. Radden J. Is this Dame Melancholy? Equating Today's Depression and past Melancholia. Philosophy, Psychiatry, & Psychology 2003; 10:37-52.

6. Szasz TS. The Manufacture of Madness: A Comparative Study of the Inquisition and the Mental Health Movement. New York: Harper and Row, Publishers Inc.; 1970.

7. Cross-national comparisons of the prevalences and correlates of mental disorders. WHO International Consortium in Psychiatric Epidemiology. Bull World Health Organ 2000; 78:413-26.

8. Kessler RC, Berglund P, Demler O, Jin R, Merikangas KR, Walters EE. Lifetime prevalence and age-of-onset distributions of DSM-IV disorders in the National Comorbidity Survey Replication. Arch Gen Psychiatry 2005; 62:593-602.

9. Kessler RC, Chiu WT, Demler O, Merikangas KR, Walters EE. Prevalence, severity, and comorbidity of 12-month DSM-IV disorders in the National Comorbidity Survey Replication. Arch Gen Psychiatry 2005; 62:617-27.

10. National Institute of Mental Health. The Numbers Count: Mental Disorders in America. [cited October 2014] <http://www.namigc.org/documents/numberscount.pdf>.

11. Demyttenaere K, Bruffaerts R, Posada-Villa J, Gasquet I, Kovess V, Lepine JP, et al. Prevalence, severity, and unmet need for treatment of mental disorders in the World Health Organization World Mental Health Surveys. JAMA 2004; 291:2581-90.

12. Torgersen S, Kringlen E, Cramer V. The prevalence of personality disorders in a community sample. Arch Gen Psychiatry 2001; 58:590-6.

13. Grant BF, Hasin DS, Stinson FS, Dawson DA, Chou SP, Ruan WJ, et al. Prevalence, correlates, and disability of personality disorders in the United States: results from the national epidemiologic survey on alcohol and related conditions. J Clin Psychiatry 2004; 65:948-58.

14. Saha S, Chant D, Welham J, McGrath J. A systematic review of the prevalence of schizophrenia. PLoS Med 2005; 2:e141.

15. Carter AS, Briggs-Gowan MJ, Davis NO. Assessment of young children's social-emotional development and psychopathology: recent advances and recommendations for practice. J Child Psychol Psychiatry 2004; 45:109-34.

16. Willcutt EG. The prevalence of DSM-IV attention-deficit/hyperactivity disorder: a meta-analytic review. Neurotherapeutics 2012; 9:490-9.

17. Fombonne E. Epidemiology of pervasive developmental disorders. Pediatr Res 2009; 65:591-8.

18. Maloney B, Sambamurti K, Zawia N, Lahiri DK. Applying epigenetics to Alzheimer's disease via the latent early-life associated regulation (LEARn) model. Curr Alzheimer Res 2012; 9:589-99.

19. Lahiri DK, Maloney B. The 'LEARn' (Latent Early-life Associated Regulation) model integrates environmental risk factors and the developmental basis of Alzheimer's disease, and proposes remedial steps. Exp Gerontol 2010; 45:291-6.

20. Lahiri DK, Maloney B, Basha MR, Ge YW, Zawia NH. How and when environmental agents and dietary factors affect the course of Alzheimer's disease: the 'LEARn' model (latent early-life associated regulation) may explain the triggering of AD. Curr Alzheimer Res 2007; 4:219-28.

21. Prince M, Bryce R, Albanese E, Wimo A, Ribeiro W, Ferri CP. The global prevalence of dementia: a systematic review and metaanalysis. Alzheimers Dement 2013; 9:63-75 e2.

22. Matthews FE, Arthur A, Barnes LE, Bond J, Jagger C, Robinson L, et al. A two-decade comparison of prevalence of dementia in

individuals aged 65 years and older from three geographical areas of England: results of the Cognitive Function and Ageing Study I and II. Lancet 2013.

23. Whitrow MJ, Moore VM, Rumbold AR, Davies MJ. Effect of supplemental folic acid in pregnancy on childhood asthma: a prospective birth cohort study. Am J Epidemiol 2009; 170:1486-93.

24. Dunstan JA, West C, McCarthy S, Metcalfe J, Meldrum S, Oddy WH, et al. The relationship between maternal folate status in pregnancy, cord blood folate levels, and allergic outcomes in early childhood. Allergy 2012; 67:50-7.

25. Stiles J, Jernigan TL. The basics of brain development. Neuropsychol Rev 2010; 20:327-48.

26. Marques AH, O'Connor TG, Roth C, Susser E, Bjorke-Monsen AL. The influence of maternal prenatal and early childhood nutrition and maternal prenatal stress on offspring immune system development and neurodevelopmental disorders. Front Neurosci 2013; 7:120.

27. Chow ML, Pramparo T, Winn ME, Barnes CC, Li HR, Weiss L, et al. Age-dependent brain gene expression and copy number anomalies in autism suggest distinct pathological processes at young versus mature ages. PLoS Genet 2012; 8:e1002592.

28. Innocenti GM, Price DJ. Exuberance in the development of cortical networks. Nat Rev Neurosci 2005; 6:955-65.

29. Hua JY, Smith SJ. Neural activity and the dynamics of central nervous system development. Nat Neurosci 2004; 7:327-32.

30. Markham JA, Greenough WT. Experience-driven brain plasticity: beyond the synapse. Neuron Glia Biol 2004; 1:351-63.

31. Szyf M, McGowan P, Meaney MJ. The social environment and the epigenome. Environ Mol Mutagen 2008; 49:46-60.

32. Sur M, Leamey CA. Development and plasticity of cortical areas and networks. Nat Rev Neurosci 2001; 2:251-62.

33. Smith J, Prior M. Temperament and stress resilience in school-age children: a within-families study. J Am Acad Child Adolesc Psychiatry 1995; 34:168-79.

34. Meaney MJ. Maternal care, gene expression, and the transmission of individual differences in stress reactivity across generations. Annu Rev Neurosci 2001; 24:1161-92.

35. Fleming AS, O'Day DH, Kraemer GW. Neurobiology of mother-infant interactions: experience and central nervous system plasticity across development and generations. Neurosci Biobehav Rev 1999; 23:673-85.

36. Pilowsky DJ, Wickramaratne P, Talati A, Tang M, Hughes CW, Garber J, et al. Children of depressed mothers 1 year after the

initiation of maternal treatment: findings from the STAR*D-Child Study. Am J Psychiatry 2008; 165:1136-47.

37. Norman RE, Byambaa M, De R, Butchart A, Scott J, Vos T. The long-term health consequences of child physical abuse, emotional abuse, and neglect: a systematic review and meta-analysis. PLoS Med 2012; 9:e1001349.

38. Montgomery SM, Bartley MJ, Wilkinson RG. Family conflict and slow growth. Arch Dis Child 1997; 77:326-30.

39. McLoyd VC. Socioeconomic disadvantage and child development. Am Psychol 1998; 53:185-204.

40. Lissau I, Sorensen TI. Parental neglect during childhood and increased risk of obesity in young adulthood. Lancet 1994; 343:324-7.

41. Holmes SJ, Robins LN. The influence of childhood disciplinary experience on the development of alcoholism and depression. J Child Psychol Psychiatry 1987; 28:399-415.

42. Choi J, Jeong B, Polcari A, Rohan ML, Teicher MH. Reduced fractional anisotropy in the visual limbic pathway of young adults witnessing domestic violence in childhood. Neuroimage 2012; 59:1071-9.

43. Russek LG, Schwartz GE. Feelings of parental caring predict health status in midlife: a 35-year follow-up of the Harvard Mastery of Stress Study. J Behav Med 1997; 20:1-13.

44. Repetti RL, Taylor SE, Seeman TE. Risky families: family social environments and the mental and physical health of offspring. Psychol Bull 2002; 128:330-66.

45. Meaney MJ. Maternal care, gene expression, and the transmission of individual differences in stress reactivity across generations. Annu Rev Neurosci 2001; 24:1161-92.

46 Szyf M, McGowan P, Meaney MJ. The social environment and the epigenome. Environ Mol Mutagen 2008; 49:46-60.

47. Weaver IC, Cervoni N, Champagne FA, D'Alessio AC, Sharma S, Seckl JR, et al. Epigenetic programming by maternal behavior. Nat Neurosci 2004; 7:847-54.

48. ibid.

49. Schaevitz LR, Berger-Sweeney JE. Gene-environment interactions and epigenetic pathways in autism: the importance of one-carbon metabolism. ILAR J 2012; 53:322-40.

50. Chen JL, Nedivi E. Neuronal structural remodeling: is it all about access? Curr Opin Neurobiol 2010; 20:557-62.

51. Schaevitz LR, Berger-Sweeney JE. Gene-environment interactions and epigenetic pathways in autism. ILAR J 2012; 53:322-40.

52. Insel TR. Rethinking schizophrenia. Nature 2010; 468:187-93.

53. Webster MJ, Knable MB, O'Grady J, Orthmann J, Weickert CS. Regional specificity of brain glucocorticoid receptor mRNA alterations in subjects with schizophrenia and mood disorders. Mol Psychiatry 2002; 7:985-94, 24.

54. McGowan PO, Sasaki A, D'Alessio AC, Dymov S, Labonte B, Szyf M, et al. Epigenetic regulation of the glucocorticoid receptor in human brain associates with childhood abuse. Nat Neurosci 2009; 12:342-8.

55. Oberlander TF, Weinberg J, Papsdorf M, Grunau R, Misri S, Devlin AM. Prenatal exposure to maternal depression, neonatal methylation of human glucocorticoid receptor gene (NR3C1) and infant cortisol stress responses. Epigenetics 2008; 3:97-106.

56. Dwivedi Y, Rizavi HS, Conley RR, Roberts RC, Tamminga CA, Pandey GN. Altered gene expression of brain-derived neurotrophic factor and receptor tyrosine kinase B in postmortem brain of suicide subjects. Arch Gen Psychiatry 2003; 60:804-15.

57. Nibuya M, Morinobu S, Duman RS. Regulation of BDNF and trkB mRNA in rat brain by chronic electroconvulsive seizure and antidepressant drug treatments. J Neurosci 1995; 15:7539-47.

58. Sun H, Kennedy PJ, Nestler EJ. Epigenetics of the depressed brain: role of histone acetylation and methylation. Neuropsychopharmacology 2013; 38:124-37.

59. Onishchenko N, Karpova N, Sabri F, Castren E, Ceccatelli S. Long-lasting depression-like behavior and epigenetic changes of BDNF gene expression induced by perinatal exposure to methylmercury. J Neurochem 2008; 106:1378-87.

60. Insel TR. Rethinking schizophrenia. Nature 2010; 468:187-93.

61. ibid.

62. Susser ES, Lin SP. Schizophrenia after prenatal exposure to the Dutch Hunger Winter of 1944-1945. Arch Gen Psychiatry 1992; 49:983-8.

63. St Clair D, Xu M, Wang P, Yu Y, Fang Y, Zhang F, et al. Rates of adult schizophrenia following prenatal exposure to the Chinese famine of 1959-1961. JAMA 2005; 294:557-62.

64. Brown AS, Derkits EJ. Prenatal infection and schizophrenia: a review of epidemiologic and translational studies. Am J Psychiatry 2010; 167:261-80.

65. Walsh T, McClellan JM, McCarthy SE, Addington AM, Pierce SB, Cooper GM, et al. Rare structural variants disrupt multiple genes in neurodevelopmental pathways in schizophrenia. Science 2008; 320:539-43.

66. Guilmatre A, Dubourg C, Mosca AL, Legallic S, Goldenberg A, Drouin-Garraud V, et al. Recurrent rearrangements in synaptic

and neurodevelopmental genes and shared biologic pathways in schizophrenia, autism, and mental retardation. Arch Gen Psychiatry 2009; 66:947-56.

67. Insel TR. Rethinking schizophrenia. Nature 2010; 468:187-93.

68. Nguyen A, Rauch TA, Pfeifer GP, Hu VW. Global methylation profiling of lymphoblastoid cell lines reveals epigenetic contributions to autism spectrum disorders and a novel autism candidate gene, RORA, whose protein product is reduced in autistic brain. FASEB J 2010; 24:3036-51.

69. Conger RD, Ge X, Elder GH, Jr., Lorenz FO, Simons RL. Economic stress, coercive family process, and developmental problems of adolescents. Child Dev 1994; 65:541-61.

70. Daskalakis NP, Bagot RC, Parker KJ, Vinkers CH, de Kloet ER. The three-hit concept of vulnerability and resilience: Toward understanding adaptation to early-life adversity outcome. Psychoneuroendocrinology 2013.

71. Saugstad LF. Infantile autism: a chronic psychosis since infancy due to synaptic pruning of the supplementary motor area. Nutr Health 2011; 20:171-82.

72. Keshavan MS, Anderson S, Pettegrew JW. Is schizophrenia due to excessive synaptic pruning in the prefrontal cortex? The Feinberg hypothesis revisited. J Psychiatr Res 1994; 28:239-65.

73. Yirmiya R, Goshen I. Immune modulation of learning, memory, neural plasticity and neurogenesis. Brain Behav Immun 2011; 25:181-213.

74. Dantzer R, Kelley KW. Twenty years of research on cytokine-induced sickness behavior. Brain Behav Immun 2007; 21:153-60.

75. Pace TW, Heim CM. A short review on the psychoneuroimmunology of posttraumatic stress disorder: from risk factors to medical comorbidities. Brain Behav Immun 2011; 25:6-13.

76. Bilbo SD, Schwarz JM. The immune system and developmental programming of brain and behavior. Front Neuroendocrinol 2012; 33:267-86.

77. ibid.

78. Brown AS, Begg MD, Gravenstein S, Schaefer CA, Wyatt RJ, Bresnahan M, et al. Serologic evidence of prenatal influenza in the etiology of schizophrenia. Arch Gen Psychiatry 2004; 61:774-80.

79. Ellis S, Mouihate A, Pittman QJ. Neonatal programming of the rat neuroimmune response: stimulus specific changes elicited by bacterial and viral mimetics. J Physiol 2006; 571:695-701.

80. Choi J, Jeong B, Polcari A, Rohan ML, Teicher MH. Reduced fractional anisotropy in the visual limbic pathway of young adults

witnessing domestic violence in childhood. Neuroimage 2012;
59:1071-9.

81. Bilbo SD, Schwarz JM. The immune system and developmental
programming of brain and behavior. Front Neuroendocrinol 2012;
33:267-86.

82. Bilbo SD, Biedenkapp JC, Der-Avakian A, Watkins LR, Rudy JW,
Maier SF. Neonatal infection-induced memory impairment after
lipopolysaccharide in adulthood is prevented via caspase-1 inhibition.
J Neurosci 2005; 25:8000-9.

83. Choy KH, de Visser YP, van den Buuse M. The effect of 'two hit'
neonatal and young-adult stress on dopaminergic modulation of
prepulse inhibition and dopamine receptor density. Br J Pharmacol
2009; 156:388-96.

84. Bilbo SD, Schwarz JM. The immune system and developmental
programming of brain and behavior. Front Neuroendocrinol 2012;
33:267-86.

85. Griffin WS, Stanley LC, Ling C, White L, MacLeod V, Perrot LJ,
et al. Brain interleukin 1 and S-100 immunoreactivity are elevated
in Down syndrome and Alzheimer disease. Proc Natl Acad Sci USA
1989; 86:7611-5.

86. Bales KR, Du Y, Holtzman D, Cordell B, Paul SM.
Neuroinflammation and Alzheimer's disease: critical roles for
cytokine/Abeta-induced glial activation, NF-kappaB, and
apolipoprotein E. Neurobiol Aging 2000; 21:427-32; discussion 51-3.

87. Sambamurti K, Granholm AC, Kindy MS, Bhat NR, Greig NH,
Lahiri DK, et al. Cholesterol and Alzheimer's disease: clinical and
experimental models suggest interactions of different genetic, dietary
and environmental risk factors. Curr Drug Targets 2004; 5:517-28.

88. Friedland RP, Fritsch T, Smyth KA, Koss E, Lerner AJ, Chen CH,
et al. Patients with Alzheimer's disease have reduced activities in
midlife compared with healthy control-group members. Proc Natl
Acad Sci USA 2001; 98:3440-5.

89. McDade TW. Early environments and the ecology of inflammation.
Proc Natl Acad Sci U S A 2012; 109 Suppl 2:17281-8.

90. Brown AS, Sourander A, Hinkka-Yli-Salomaki S, McKeague IW,
Sundvall J, Surcel HM. Elevated maternal C-reactive protein and
autism in a national birth cohort. Mol Psychiatry 2013.

91. Lahiri DK, Maloney B. The 'LEARn' (latent early-life associated
regulation) model: an epigenetic pathway linking metabolic and
cognitive disorders. J Alzheimers Dis 2012; 30 Suppl 2:S15-30.

92. Leitner Y, Fattal-Valevski A, Geva R, Eshel R, Toledano-Alhadef H,
Rotstein M, et al. Neurodevelopmental outcome of children with

intrauterine growth retardation: a longitudinal, 10-year prospective study. J Child Neurol 2007; 22:580-7.

93. Brown AS, Susser ES. Prenatal nutritional deficiency and risk of adult schizophrenia. Schizophr Bull 2008; 34:1054-63.

94. Marques AH, O'Connor TG, Roth C, Susser E, Bjorke-Monsen AL. The influence of maternal prenatal and early childhood nutrition and maternal prenatal stress on offspring immune system development and neurodevelopmental disorders. Front Neurosci 2013; 7:120.

95. Latzin P, Frey U, Armann J, Kieninger E, Fuchs O, Roosli M, et al. Exposure to moderate air pollution during late pregnancy and cord blood cytokine secretion in healthy neonates. PLoS One 2011; 6:e23130.

96. Liu J, Ballaney M, Al-alem U, Quan C, Jin X, Perera F, et al. Combined inhaled diesel exhaust particles and allergen exposure alter methylation of T helper genes and IgE production in vivo. Toxicol Sci 2008; 102:76-81.

97. Bolton JL, Smith SH, Huff NC, Gilmour MI, Foster WM, Auten RL, et al. Prenatal air pollution exposure induces neuroinflammation and predisposes offspring to weight gain in adulthood in a sex-specific manner. FASEB J 2012; 26:4743-54.

98. Park HY, Hertz-Picciotto I, Sovcikova E, Kocan A, Drobna B, Trnovec T. Neurodevelopmental toxicity of prenatal polychlorinated biphenyls (PCBs) by chemical structure and activity: a birth cohort study. Environ Health 2010; 9:51.

99. Walkowiak J, Wiener JA, Fastabend A, Heinzow B, Kramer U, Schmidt E, et al. Environmental exposure to polychlorinated biphenyls and quality of the home environment: effects on psychodevelopment in early childhood. Lancet 2001; 358:1602-7.

100. Dietert RR, Dietert JM. Potential for early-life immune insult including developmental immunotoxicity in autism and autism spectrum disorders: focus on critical windows of immune vulnerability. J Toxicol Environ Health B Crit Rev 2008; 11:660-80.

101. Nieuwenhuijsen MJ, Martinez D, Grellier J, Bennett J, Best N, Iszatt N, et al. Chlorination disinfection by-products in drinking water and congenital anomalies: review and meta-analyses. Environ Health Perspect 2009; 117:1486-93.

102. Mayer EA. Gut feelings: the emerging biology of gut-brain communication. Nat Rev Neurosci 2011; 12:453-66.

103. Bercik P, Collins SM, Verdu EF. Microbes and the gut-brain axis. Neurogastroenterol Motil 2012; 24:405-13.

104. Dinan TG, Cryan JF. Regulation of the stress response by the gut microbiota: Implications for psychoneuroendocrinology. Psychoneuroendocrinology 2012; 37:1369-78.

105. Sun H, Kennedy PJ, Nestler EJ. Epigenetics of the depressed brain: role of histone acetylation and methylation. Neuropsychopharmacology 2013; 38:124-37.

106. Sudo N, Sawamura S, Tanaka K, Aiba Y, Kubo C, Koga Y. The requirement of intestinal bacterial flora for the development of an IgE production system fully susceptible to oral tolerance induction. J Immunol 1997; 159:1739-45.

107. Sudo N, Chida Y, Aiba Y, Sonoda J, Oyama N, Yu XN, et al. Postnatal microbial colonization programs the hypothalamic-pituitary-adrenal system for stress response in mice. J Physiol 2004; 558:263-75.

108. St Clair D, Xu M, Wang P, Yu Y, Fang Y, Zhang F, et al. Rates of adult schizophrenia following prenatal exposure to the Chinese famine of 1959-1961. JAMA 2005; 294:557-62.

109. Bercik P, Denou E, Collins J, Jackson W, Lu J, Jury J, et al. The intestinal microbiota affect central levels of brain-derived neurotropic factor and behavior in mice. Gastroenterology 2011; 141:599-609, e1-3.

110. Logan AC, Katzman M. Major depressive disorder: probiotics may be an adjuvant therapy. Medical Hypotheses 2005; 64:533-8.

111. Tillisch K, Labus J, Kilpatrick L, Jiang Z, Stains J, Ebrat B, et al. Consumption of fermented milk product with probiotic modulates brain activity. Gastroenterology 2013; 144:1394-401, 401 e1-4.

112. Critchfield JW, van Hemert S, Ash M, Mulder L, Ashwood P. The potential role of probiotics in the management of childhood autism spectrum disorders. Gastroenterol Res Pract 2011; 2011:161358.

113. Emanuele E, Orsi P, Boso M, Broglia D, Brondino N, Barale F, et al. Low-grade endotoxemia in patients with severe autism. Neurosci Lett 2010; 471:162-5.

114. de Magistris L, Familiari V, Pascotto A, Sapone A, Frolli A, Iardino P, et al. Alterations of the intestinal barrier in patients with autism spectrum disorders and in their first-degree relatives. J Pediatr Gastroenterol Nutr 2010; 51:418-24.

115. Takano T, Nakamura K, Watanabe M. Urban residential environments and senior citizens' longevity in megacity areas: the importance of walkable green spaces. J Epidemiol Community Health 2002; 56:913-8.

116. Sharp D. Giving people more green space. J Urban Health 2007; 84:3-4.

117. Maas J, Verheij RA, Groenewegen PP, de Vries S, Spreeuwenberg P. Green space, urbanity, and health: how strong is the relation? J Epidemiol Community Health 2006; 60:587-92.

118. Villeneuve PJ, Jerrett M, Su JG, Burnett RT, Chen H, Wheeler AJ, et al. A cohort study relating urban green space with mortality in Ontario, Canada. Environ Res 2012; 115:51-8.

119. Selhub EM, Logan AC. Your Brain on Nature: The Science of Nature's Influence on Your Health, Happiness and Vitality. Mississauga, Canada: Wiley; 2012.

120. Lee J, Park BJ, Tsunetsugu Y, Ohira T, Kagawa T, Miyazaki Y. Effect of forest bathing on physiological and psychological responses in young Japanese male subjects. Public Health 2011; 125:93-100.

121. Park BJ, Tsunetsugu Y, Kasetani T, Hirano H, Kagawa T, Sato M, et al. Physiological effects of Shinrin-yoku (taking in the atmosphere of the forest)–using salivary cortisol and cerebral activity as indicators. J Physiol Anthropol 2007; 26:123-8.

122. Park BJ, Tsunetsugu Y, Kasetani T, Kagawa T, Miyazaki Y. The physiological effects of Shinrin-yoku (taking in the forest atmosphere or forest bathing): evidence from field experiments in 24 forests across Japan. Environ Health Prev Med 2010; 15:18-26.

123. Kim GW, Jeong GW, Kim TH, Baek HS, Oh SK, Kang HK, et al. Functional neuroanatomy associated with natural and urban scenic views in the human brain: 3.0T functional MR imaging. Korean J Radiol 2010; 11:507-13.

124. Ulrich RS. View through a window may influence recovery from surgery. Science 1984; 224:420-3.

125. Berman MG, Jonides J, Kaplan S. The cognitive benefits of interacting with nature. Psychol Sci 2008; 19:1207-12.

126. Daly J, Burchett M, Torpy F. Plants in the classroom can improve student performance. 2010 Plants in the classroom can improve student performance, <http://www.interiorplantscape.asn.au/plants-in-schools-full-report>.

127. Matsuoka RH. Student Performance and High School Landscapes: Examining the Links. Landscape and Urban Planning 2010; 97(4):273–82.

128. Taylor AF, Kuo FE. Children with attention deficits concentrate better after walk in the park. J Atten Disord 2009; 12:402-9.

Chapter 8: The early-life influences on our musculoskeletal and locomotor systems (bones, muscles and joints)

1. Helmick CG, Felson DT, Lawrence RC, Gabriel S, Hirsch R, Kwoh CK, et al. Estimates of the prevalence of arthritis and other rheumatic conditions in the United States. Part I. Arthritis Rheum 2008; 58:15-25.

2. Fransen M, Bridgett L, March L, Hoy D, Penserga E, Brooks P. The epidemiology of osteoarthritis in Asia. Int J Rheum Dis 2011; 14:113-21.

3. Riggs BL, Melton LJ, 3rd. The worldwide problem of osteoporosis: insights afforded by epidemiology. Bone 1995; 17:505S-11S.

4. Gullberg B, Johnell O, Kanis JA. World-wide projections for hip fracture. Osteoporos Int 1997; 7:407-13.

5. Cooper C, Campion G, Melton LJ, 3rd. Hip fractures in the elderly: a world-wide projection. Osteoporos Int 1992; 2:285-9.

6. Holroyd C, Harvey N, Dennison E, Cooper C. Epigenetic influences in the developmental origins of osteoporosis. Osteoporos Int 2012; 23:401-10.

7. Hui SL, Slemenda CW, Johnston CC, Jr. The contribution of bone loss to postmenopausal osteoporosis. Osteoporos Int 1990; 1:30-4.

8. Hernandez CJ, Beaupre GS, Carter DR. A theoretical analysis of the relative influences of peak BMD, age-related bone loss and menopause on the development of osteoporosis. Osteoporos Int 2003; 14:843-7.

9. Cooper C, Eriksson JG, Forsen T, Osmond C, Tuomilehto J, Barker DJ. Maternal height, childhood growth and risk of hip fracture in later life: a longitudinal study. Osteoporos Int 2001; 12:623-9.

10. Cooper C, Fall C, Egger P, Hobbs R, Eastell R, Barker D. Growth in infancy and bone mass in later life. Ann Rheum Dis 1997; 56:17-21.

11. Dennison EM, Syddall HE, Sayer AA, Gilbody HJ, Cooper C. Birth weight and weight at 1 year are independent determinants of bone mass in the seventh decade: the Hertfordshire cohort study. Pediatr Res 2005; 57:582-6.

12. Oliver H, Jameson KA, Sayer AA, Cooper C, Dennison EM. Growth in early life predicts bone strength in late adulthood: the Hertfordshire Cohort Study. Bone 2007; 41:400-5.

13. Javaid MK, Lekamwasam S, Clark J, Dennison EM, Syddall HE, Loveridge N, et al. Infant growth influences proximal femoral geometry in adulthood. J Bone Miner Res 2006; 21:508-12.

14. Done SL. Fetal and neonatal bone health: update on bone growth and manifestations in health and disease. Pediatr Radiol 2012; 42 Suppl 1:S158-76.

15. Javaid MK, Crozier SR, Harvey NC, Gale CR, Dennison EM, Boucher BJ, et al. Maternal vitamin D status during pregnancy and childhood bone mass at age 9 years: a longitudinal study. Lancet 2006; 367:36-43.

16. Namgung R, Tsang RC, Lee C, Han DG, Ho ML, Sierra RI. Low total body bone mineral content and high bone resorption in Korean

winter-born versus summer-born newborn infants. J Pediatr 1998; 132:421-5.

17. Abrahamsen B, Heitmann BL, Eiken PA. Season of birth and the risk of hip fracture in danish men and women aged 65+. Front Endocrinol (Lausanne) 2012; 3:2.

18. Ganpule A, Yajnik CS, Fall CH, Rao S, Fisher DJ, Kanade A, et al. Bone mass in Indian children—relationships to maternal nutritional status and diet during pregnancy: the Pune Maternal Nutrition Study. J Clin Endocrinol Metab 2006; 91:2994-3001.

19. Koo WW, Walters JC, Esterlitz J, Levine RJ, Bush AJ, Sibai B. Maternal calcium supplementation and fetal bone mineralization. Obstet Gynecol 1999; 94:577-82.

20. Martin R, Harvey NC, Crozier SR, Poole JR, Javaid MK, Dennison EM, et al. Placental calcium transporter (PMCA3) gene expression predicts intrauterine bone mineral accrual. Bone 2007; 40:1203-8.

21. Javaid MK, Godfrey KM, Taylor P, Robinson SM, Crozier SR, Dennison EM, et al. Umbilical cord leptin predicts neonatal bone mass. Calcif Tissue Int 2005; 76:341-7.

22. Dennison E, Hindmarsh P, Fall C, Kellingray S, Barker D, Phillips D, et al. Profiles of endogenous circulating cortisol and bone mineral density in healthy elderly men. J Clin Endocrinol Metab 1999; 84:3058-63.

23. Dennison EM, Syddall HE, Rodriguez S, Voropanov A, Day IN, Cooper C. Polymorphism in the growth hormone gene, weight in infancy, and adult bone mass. J Clin Endocrinol Metab 2004; 89:4898-903.

24. Fall C, Hindmarsh P, Dennison E, Kellingray S, Barker D, Cooper C. Programming of growth hormone secretion and bone mineral density in elderly men: a hypothesis. J Clin Endocrinol Metab 1998; 83:135-9.

25. Weaver IC, Cervoni N, Champagne FA, D'Alessio AC, Sharma S, Seckl JR, et al. Epigenetic programming by maternal behavior. Nat Neurosci 2004; 7:847-54.

26. Mehta G, Roach HI, Langley-Evans S, Taylor P, Reading I, Oreffo RO, et al. Intrauterine exposure to a maternal low protein diet reduces adult bone mass and alters growth plate morphology in rats. Calcif Tissue Int 2002; 71:493-8.

27. Cole ZA, Gale CR, Javaid MK, Robinson SM, Law C, Boucher BJ, et al. Maternal dietary patterns during pregnancy and childhood bone mass: a longitudinal study. J Bone Miner Res 2009; 24:663-8.

28. Tucker KL. Osteoporosis prevention and nutrition. Curr Osteoporos Rep 2009; 7:111-7.

29. El-Makawy AI, Girgis SM, Khalil WK. Developmental and genetic toxicity of stannous chloride in mouse dams and fetuses. Mutat Res 2008; 657:105-10.

30. Harvey NC, Poole JR, Javaid MK, Dennison EM, Robinson S, Inskip HM, et al. Parental determinants of neonatal body composition. J Clin Endocrinol Metab 2007; 92:523-6.

31. Harvey NC, Javaid MK, Poole JR, Taylor P, Robinson SM, Inskip HM, et al. Paternal skeletal size predicts intrauterine bone mineral accrual. J Clin Endocrinol Metab 2008; 93:1676-81.

32. Blencowe H, Cousens S, Oestergaard MZ, Chou D, Moller AB, Narwal R, et al. National, regional, and worldwide estimates of preterm birth rates in the year 2010 with time trends since 1990 for selected countries: a systematic analysis and implications. Lancet 2012; 379:2162-72.

33. Backstrom MC, Maki R, Kuusela AL, Sievanen H, Koivisto AM, Koskinen M, et al. The long-term effect of early mineral, vitamin D, and breast milk intake on bone mineral status in 9- to 11-year-old children born prematurely. J Pediatr Gastroenterol Nutr 1999; 29:575-82.

34. Fewtrell M. Early nutritional predictors of long-term bone health in preterm infants. Curr Opin Clin Nutr Metab Care 2011; 14:297-301.

35. Kuhn T, Kroke A, Remer T, Schonau E, Buyken AE. Is breastfeeding related to bone properties? A longitudinal analysis of associations between breastfeeding duration and pQCT parameters in children and adolescents. Matern Child Nutr 2012.

36. Fewtrell MS, Prentice A, Jones SC, Bishop NJ, Stirling D, Buffenstein R, et al. Bone mineralization and turnover in preterm infants at 8-12 years of age: the effect of early diet. J Bone Miner Res 1999; 14:810-20.

37. Weiler HA, Yuen CK, Seshia MM. Growth and bone mineralization of young adults weighing less than 1500 g at birth. Early Hum Dev 2002; 67:101-12.

38. Parkinson JR, Hyde MJ, Gale C, Santhakumaran S, Modi N. Preterm birth and the metabolic syndrome in adult life: a systematic review and meta-analysis. Pediatrics 2013; 131:e1240-63.

39. Lewandowski AJ, Augustine D, Lamata P, Davis EF, Lazdam M, Francis J, et al. Preterm heart in adult life: cardiovascular magnetic resonance reveals distinct differences in left ventricular mass, geometry, and function. Circulation 2013; 127:197-206.

40. Johnson MJ, Wootton SA, Leaf AA, Jackson AA. Preterm birth and body composition at term equivalent age: a systematic review and meta-analysis. Pediatrics 2012; 130:e640-9.

41. Cooper C, Eriksson JG, Forsen T, Osmond C, Tuomilehto J, Barker DJ. Maternal height, childhood growth and risk of hip fracture in later life: a longitudinal study. Osteoporos Int 2001; 12:623-9.
42. Wyshak G. Teenaged girls, carbonated beverage consumption, and bone fractures. Arch Pediatr Adolesc Med 2000; 154:610-3.
43. Vartanian LR, Schwartz MB, Brownell KD. Effects of soft drink consumption on nutrition and health: a systematic review and meta-analysis. Am J Public Health 2007; 97:667-75.
44. ibid.
45. Ka K, Rousseau MC, Lambert M, O'Loughlin J, Henderson M, Tremblay A, et al. Association between Lean and Fat Mass and Indicators of Bone Health in Prepubertal Caucasian Children. Horm Res Paediatr 2013: 80(3):154-62 from PubMed.
46. Gunter KB, Almstedt HC, Janz KF. Physical activity in childhood. Exerc Sport Sci Rev 2012; 40:13-21.
47. Ka K, Rousseau MC, Lambert M, O'Loughlin J, Henderson M, Tremblay A, et al. Association between Lean and Fat Mass and Indicators of Bone Health in Prepubertal Caucasian Children. Horm Res Paediatr 2013.
48. Hind K, Burrows M. Weight-bearing exercise and bone mineral accrual in children and adolescents: a review of controlled trials. Bone 2007; 40:14-27.
49. Gunter K, Baxter-Jones AD, Mirwald RL, Almstedt H, Fuller A, Durski S, et al. Jump starting skeletal health: a 4-year longitudinal study assessing the effects of jumping on skeletal development in pre and circum pubertal children. Bone 2008; 42:710-8.
50. Gunter K, Baxter-Jones AD, Mirwald RL, Almstedt H, Fuchs RK, Durski S, et al. Impact exercise increases BMC during growth: an 8-year longitudinal study. J Bone Miner Res 2008; 23:986-93.
51. Caine DJ, Golightly YM. Osteoarthritis as an outcome of paediatric sport: an epidemiological perspective. Br J Sports Med 2011; 45:298-303.
52. Janz KF, Gilmore JM, Levy SM, Letuchy EM, Burns TL, Beck TJ. Physical activity and femoral neck bone strength during childhood: the Iowa Bone Development Study. Bone 2007; 41:216-22.
53. Booth FW, Roberts CK, Laye MJ. Lack of exercise is a major cause of chronic diseases. Compr Physiol 2012; 2:1143-211.
54. FDI World Dental Federation. Oral health and the United Nations Political Declaration on NCDs A guide to advocacy. 2012 [cited October 2014] < http://www.fdiworldental.org/media/9465/oral_health_and_un_political_dec_on_ncds.pdf>.

55. Caplan DJ, Chasen JB, Krall EA, Cai J, Kang S, Garcia RI, et al. Lesions of endodontic origin and risk of coronary heart disease. J Dent Res 2006; 85:996-1000.

56. Armelagos GJ, Goodman AH, Harper KN, Blakey ML. Enamel Hypoplasia and Earlier Mortality: Bioarchaeological Support for the Barker Hypothesis. Evolutionary Anthropology 2009; 18:261-71.

57. White TD. Early hominid enamel hypoplasia. Am J Phys Anthropol 1978; 49:79-83.

58. Emory University. Ancient human teeth show that stress early in development can shorten life span. ScienceDaily, 5 February 2010.

59. Armelagos GJ, Goodman AH, Harper KN, Blakey ML. Enamel Hypoplasia and Earlier Mortality: Bioarchaeological Support for the Barker Hypothesis. Evolutionary Anthropology 2009; 18:261-71.

60. ibid.

61. Lappin K, Kealey D, Cosgrove A, Graham K. Does low birthweight predispose to Perthes' disease? Perthes' disease in twins. J Pediatr Orthop B 2003; 12:307-10.

62. Perry DC, Hall AJ. The epidemiology and etiology of Perthes disease. Orthop Clin North Am 2011; 42:279-83, v.

63. Perry DC, Bruce CE, Pope D, Dangerfield P, Platt MJ, Hall AJ. Legg-Calve-Perthes disease in the UK: geographic and temporal trends in incidence reflecting differences in degree of deprivation in childhood. Arthritis Rheum 2012; 64:1673-9.

64. Straker LM, Coleman J, Skoss R, Maslen BA, Burgess-Limerick R, Pollock CM. A comparison of posture and muscle activity during tablet computer, desktop computer and paper use by young children. Ergonomics 2008; 51:540-55.

65. Australian Bureau of Statistics. Children's participation in cultural and leisure activities. 2011, Australia. ABS catalogue no. 4102.0.

66. Straker L, Harris C. Survey of physical ergonomics issues associated with school children's use of laptop computers. . International Journal of Industrial Ergonomics 2000; 26:337-46.

67. Hakala PT, Rimpela AH, Saarni LA, Salminen JJ. Frequent computer-related activities increase the risk of neck-shoulder and low back pain in adolescents. Eur J Public Health 2006; 16:536-41.

68. Sommerich CM, Ward R, Sikdar K, Payne J, Herman L. A survey of high school students with ubiquitous access to tablet PCs. Ergonomics 2007; 50:706-27.

69. Straker LM, O'Sullivan PB, Smith A, Perry M. Computer use and habitual spinal posture in Australian adolescents. Public Health Rep 2007; 122:634-43.

70. Straker LM, Coleman J, Skoss R, Maslen BA, Burgess-Limerick R, Pollock CM. A comparison of posture and muscle activity during tablet computer, desktop computer and paper use by young children. Ergonomics 2008; 51:540-55.

71. van der Ploeg HP, Chey T, Korda RJ, Banks E, Bauman A. Sitting time and all-cause mortality risk in 222 497 Australian adults. Arch Intern Med 2012; 172:494-500.

72. Chau JY, Grunseit A, Midthjell K, Holmen J, Holmen TL, Bauman AE, et al. Sedentary behaviour and risk of mortality from all-causes and cardiometabolic diseases in adults: evidence from the HUNT3 population cohort. Br J Sports Med 2013.

73. Tremblay MS, LeBlanc AG, Kho ME, Saunders TJ, Larouche R, Colley RC, et al. Systematic review of sedentary behaviour and health indicators in school-aged children and youth. Int J Behav Nutr Phys Act 2011; 8:98.

74. Manners PJ, Bower C. Worldwide prevalence of juvenile arthritis – why does it vary so much? J Rheumatol 2002; 29:1520-30.

75. Colebatch AN, Edwards CJ. The influence of early life factors on the risk of developing rheumatoid arthritis. Clin Exp Immunol 2011; 163:11-6.

76. Sugiyama D, Nishimura K, Tamaki K, Tsuji G, Nakazawa T, Morinobu A, et al. Impact of smoking as a risk factor for developing rheumatoid arthritis: a meta-analysis of observational studies. Ann Rheum Dis 2010; 69:70-81.

77. Linn-Rasker SP, van der Helm-van Mil AH, van Gaalen FA, Kloppenburg M, de Vries RR, le Cessie S, et al. Smoking is a risk factor for anti-CCP antibodies only in rheumatoid arthritis patients who carry HLA-DRB1 shared epitope alleles. Ann Rheum Dis 2006; 65:366-71.

78. Klareskog L, Stolt P, Lundberg K, Kallberg H, Bengtsson C, Grunewald J, et al. A new model for an etiology of rheumatoid arthritis: smoking may trigger HLA-DR (shared epitope)-restricted immune reactions to autoantigens modified by citrullination. Arthritis Rheum 2006; 54:38-46.

79. Jaakkola JJ, Gissler M. Maternal smoking in pregnancy as a determinant of rheumatoid arthritis and other inflammatory polyarthropathies during the first 7 years of life. Int J Epidemiol 2005; 34:664-71.

80. Carlens C, Jacobsson L, Brandt L, Cnattingius S, Stephansson O, Askling J. Perinatal characteristics, early life infections and later risk of rheumatoid arthritis and juvenile idiopathic arthritis. Ann Rheum Dis 2009; 68:1159-64.

81. Edwards CJ, Goswami R, Goswami P, Syddall H, Dennison EM, Arden NK, et al. Growth and infectious exposure during infancy and the risk of rheumatoid factor in adult life. Ann Rheum Dis 2006; 65:401-4.

82. Merlino LA, Curtis J, Mikuls TR, Cerhan JR, Criswell LA, Saag KG. Vitamin D intake is inversely associated with rheumatoid arthritis: results from the Iowa Women's Health Study. Arthritis Rheum 2004; 50:72-7.

83. Cutolo M, Otsa K, Paolino S, Yprus M, Veldi T, Seriolo B. Vitamin D involvement in rheumatoid arthritis and systemic lupus erythaematosus. Ann Rheum Dis 2009; 68:446-7.

84. Costenbader KH, Feskanich D, Holmes M, Karlson EW, Benito-Garcia E. Vitamin D intake and risks of systemic lupus erythematosus and rheumatoid arthritis in women. Ann Rheum Dis 2008; 67:530-5.

85. Hypponen E, Laara E, Reunanen A, Jarvelin MR, Virtanen SM. Intake of vitamin D and risk of type 1 diabetes: a birth-cohort study. Lancet 2001; 358:1500-3.

86. Lahiri M, Morgan C, Symmons DP, Bruce IN. Modifiable risk factors for RA: prevention, better than cure? Rheumatology (Oxford) 2012; 51:499-512.

87. Shapiro JA, Koepsell TD, Voigt LF, Dugowson CE, Kestin M, Nelson JL. Diet and rheumatoid arthritis in women: a possible protective effect of fish consumption. Epidemiology 1996; 7:256-63.

88. Pedersen M, Stripp C, Klarlund M, Olsen SF, Tjonneland AM, Frisch M. Diet and risk of rheumatoid arthritis in a prospective cohort. J Rheumatol 2005; 32:1249-52.

89. Prescott SL, Calder PC. N-3 polyunsaturated fatty acids and allergic disease. Curr Opin Clin Nutr Metab Care 2004; 7:123–9.

90. Dunstan J, Mori TA, Barden A, Beilin LJ, Taylor A, Holt PG, et al. Fish oil supplementation in pregnancy modifies neonatal allergen-specific immune responses and clinical outcomes in infants at high risk of atopy: a randomised controlled trial. J Allergy Clin Immunol 2003; 112:1178-84.

91. Palmer DJ, Sullivan T, Gold MS, Prescott SL, Heddle R, Gibson RA, et al. Effect of n-3 long chain polyunsaturated fatty acid supplementation in pregnancy on infants' allergies in first year of life: randomised controlled trial. BMJ 2012; 344:e184.

92. Jacobsson LT, Jacobsson ME, Askling J, Knowler WC. Perinatal characteristics and risk of rheumatoid arthritis. BMJ 2003; 326:1068-9.

93. Simard JF, Costenbader KH, Hernan MA, Liang MH, Mittleman MA, Karlson EW. Early life factors and adult-onset rheumatoid arthritis. J Rheumatol 2010; 37:32-7.

94. Griffin TM, Fermor B, Huebner JL, Kraus VB, Rodriguiz RM, Wetsel WC, et al. Diet-induced obesity differentially regulates behavioral, biomechanical, and molecular risk factors for osteoarthritis in mice. Arthritis Res Ther 2010; 12:R130.

95. Colebatch AN, Edwards CJ. The influence of early life factors on the risk of developing rheumatoid arthritis. Clin Exp Immunol 2011; 163:11-6.

96. Griffin TM, Fermor B, Huebner JL, Kraus VB, Rodriguiz RM, Wetsel WC, et al. Diet-induced obesity differentially regulates behavioral, biomechanical, and molecular risk factors for osteoarthritis in mice. Arthritis Res Ther 2010; 12:R130.'

97. Rai MF, Sandell LJ. Inflammatory mediators: tracing links between obesity and osteoarthritis. Crit Rev Eukaryot Gene Expr 2011; 21:131-42.

98. Bultink IE, Lems WF. Osteoarthritis and osteoporosis: what is the overlap? Curr Rheumatol Rep 2013; 15:328.

99. Schett G, Kiechl S, Weger S, Pederiva A, Mayr A, Petrangeli M, et al. High-sensitivity C-reactive protein and risk of nontraumatic fractures in the Bruneck study. Arch Intern Med 2006; 166:2495-501.

100. McDade TW. Early environments and the ecology of inflammation. Proc Natl Acad Sci USA 2012; 109 Suppl 2:17281-8.

101. Lane NE. Exercise: a cause of osteoarthritis. J Rheumatol Suppl 1995; 43:3-6.

102. Buckwalter JA. Osteoarthritis and articular cartilage use, disuse, and abuse: experimental studies. J Rheumatol Suppl 1995; 43:13-5.

103. Felson DT, Niu J, Clancy M, Sack B, Aliabadi P, Zhang Y. Effect of recreational physical activities on the development of knee osteoarthritis in older adults of different weights: the Framingham Study. Arthritis Rheum 2007; 57:6-12.

104. Non-Communicable Diseases (NCDs): a priority for women's health and development. The NCD Alliance. 2011. [Cited Octer 2014] < http://www.who.int/pmnch/topics/maternal/2011_women_ncd_report.pdf.pdf>.

105. Women and Health: Today's evidence, tomorrow's agenda. World Health Organization, Geneva. 2009. [Cited October 2014] < http://www.who.int/gender/women_health_report/full_report_20091104_en.pdf>.

106. Wen LM, Baur LA, Simpson JM, Rissel C, Wardle K, Flood VM. Effectiveness of home based early intervention on children's BMI at age 2: randomised controlled trial. BMJ 2012; 344:e3732.

107. Hesketh KD, Campbell KJ. Interventions to prevent obesity in 0-5 year olds: an updated systematic review of the literature. Obesity (Silver Spring) 2010; 18 Suppl 1:S27-35.
108. Krishnaswami J, Martinson M, Wakimoto P, Anglemeyer A. Community-engaged interventions on diet, activity, and weight outcomes in U.S. schools: a systematic review. Am J Prev Med 2012; 43:81-91.

Chapter 9: The developmental origins of cancer

1. Jemal A, Bray F, Center MM, Ferlay J, Ward E, Forman D. Global cancer statistics. CA Cancer J Clin 2011; 61:69-90.
2. Kaatsch P. Epidemiology of childhood cancer. Cancer Treat Rev 2010; 36:277-85.
3. Ward EM, Thun MJ, Hannan LM, Jemal A. Interpreting cancer trends. Ann N Y Acad Sci 2006; 1076:29-53.
4. Lane DP. Cancer. p53, guardian of the genome. Nature 1992; 358:15-6.
5. Efeyan A, Serrano M. p53: guardian of the genome and policeman of the oncogenes. Cell Cycle 2007; 6:1006-10.
6. Koshland DE, Jr. The molecule of the year. Science 1992; 258:1861.
7. Vogelstein B. Cancer. A deadly inheritance. Nature 1990; 348:681-2.
8. Greenblatt MS, Bennett WP, Hollstein M, Harris CC. Mutations in the p53 tumor suppressor gene: clues to cancer etiology and molecular pathogenesis. Cancer Res 1994; 54:4855-78.
9. Merlo LM, Pepper JW, Reid BJ, Maley CC. Cancer as an evolutionary and ecological process. Nat Rev Cancer 2006; 6:924-35.
10. Reiche EM, Nunes SO, Morimoto HK. Stress, depression, the immune system, and cancer. Lancet Oncol 2004; 5:617-25.
11. Rowley JD. Letter: A new consistent chromosomal abnormality in chronic myelogenous leukaemia identified by quinacrine fluorescence and Giemsa staining. Nature 1973; 243:290-3.
12. Schnekenburger M, Diederich M. Epigenetics Offer New Horizons for Colorectal Cancer Prevention. Curr Colorectal Cancer Rep 2012; 8:66-81.
13. Balaguer F, Link A, Lozano JJ, Cuatrecasas M, Nagasaka T, Boland CR, et al. Epigenetic silencing of miR-137 is an early event in colorectal carcinogenesis. Cancer Res 2010; 70:6609-18.
14. Jacinto FV, Esteller M. Mutator pathways unleashed by epigenetic silencing in human cancer. Mutagenesis 2007; 22:247-53.
15. Lahtz C, Pfeifer GP. Epigenetic changes of DNA repair genes in cancer. J Mol Cell Biol 2011; 3:51-8.

16. Virani S, Colacino JA, Kim JH, Rozek LS. Cancer epigenetics: a brief review. ILAR J 2012; 53:359-69.
17. Roukos DH. Genome-wide association studies: how predictable is a person's cancer risk? Expert Rev Anticancer Ther 2009; 9:389-92.
18. ibid.
19. Vogelstein B. Cancer. A deadly inheritance. Nature 1990; 348:681-2.
20. Ho SM, Johnson A, Tarapore P, Janakiram V, Zhang X, Leung YK. Environmental epigenetics and its implication on disease risk and health outcomes. ILAR J 2012; 53:289-305.
21. Sonnenschein C, Wadia PR, Rubin BS, Soto AM. Cancer as development gone awry: the case for bisphenol-A as a carcinogen. J Dev Origins Health Disease 2011; 2:9-16.
22. Walker CL, Ho SM. Developmental reprogramming of cancer susceptibility. Nat Rev Cancer 2012; 12:479-86.
23. ibid.
24. Herbst AL, Ulfelder H, Poskanzer DC. Adenocarcinoma of the vagina. Association of maternal stilbestrol therapy with tumor appearance in young women. N Engl J Med 1971; 284:878-81.
25. Mittendorf R. Teratogen update: carcinogenesis and teratogenesis associated with exposure to diethylstilbestrol (DES) in utero. Teratology 1995; 51:435-45.
26. McLachlan JA, Newbold RR, Bullock BC. Long-term effects on the female mouse genital tract associated with prenatal exposure to diethylstilbestrol. Cancer Res 1980; 40:3988-99.
27. McLachlan JA, Newbold RR, Bullock B. Reproductive tract lesions in male mice exposed prenatally to diethylstilbestrol. Science 1975; 190:991-2.
28. Palmer JR, Wise LA, Hatch EE, Troisi R, Titus-Ernstoff L, Strohsnitter W, et al. Prenatal diethylstilbestrol exposure and risk of breast cancer. Cancer Epidemiol Biomarkers Prev 2006; 15:1509-14.
29. Skakkebaek NE, Rajpert-De Meyts E, Jorgensen N, Carlsen E, Petersen PM, Giwercman A, et al. Germ cell cancer and disorders of spermatogenesis: an environmental connection? APMIS 1998; 106:3-11; discussion 2.
30. ibid.
31. Sonnenschein C, Wadia PR, Rubin BS, Soto AM. Cancer as development gone awry. J Dev Origins Health Disease 2011; 2:9–16.
32. Durando M, Kass L, Piva J, Sonnenschein C, Soto AM, Luque EH, et al. Prenatal bisphenol A exposure induces preneoplastic lesions in the mammary gland in Wistar rats. Environ Health Perspect 2007; 115:80-6.

33. Murray TJ, Maffini MV, Ucci AA, Sonnenschein C, Soto AM. Induction of mammary gland ductal hyperplasias and carcinoma in situ following fetal bisphenol A exposure. Reprod Toxicol 2007; 23:383-90.

34. Ho SM, Tang WY, Belmonte de Frausto J, Prins GS. Developmental exposure to estradiol and bisphenol A increases susceptibility to prostate carcinogenesis and epigenetically regulates phosphodiesterase type 4 variant 4. Cancer Res 2006; 66:5624-32.

35. Castano-Vinyals G, Carrasco E, Lorente JA, Sabate Y, Cirac-Claveras J, Pollan M, et al. Anogenital distance and the risk of prostate cancer. BJU Int 2012; 110:E707-10.

36. Soto AM, Sonnenschein C. Environmental causes of cancer: endocrine disruptors as carcinogens. Nat Rev Endocrinol 2010; 6:363-70.

37. Manikkam M, Tracey R, Guerrero-Bosagna C, Skinner MK. Plastics derived endocrine disruptors (BPA, DEHP and DBP) induce epigenetic transgenerational inheritance of obesity, reproductive disease and sperm epimutations. PLoS One 2013; 8:e55387.

38. Swan SH, Main KM, Liu F, Stewart SL, Kruse RL, Calafat AM, et al. Decrease in anogenital distance among male infants with prenatal phthalate exposure. Environ Health Perspect 2005; 113:1056-61.

39. Tyl RW, Myers CB, Marr MC, Fail PA, Seely JC, Brine DR, et al. Reproductive toxicity evaluation of dietary butyl benzyl phthalate (BBP) in rats. Reprod Toxicol 2004; 18:241-64.

40. Lund L, Engebjerg MC, Pedersen L, Ehrenstein V, Norgaard M, Sorensen HT. Prevalence of hypospadias in Danish boys: a longitudinal study, 1977-2005. Eur Urol 2009; 55:1022-6.

41. Hsieh MH, Eisenberg ML, Hittelman AB, Wilson JM, Tasian GE, Baskin LS. Caucasian male infants and boys with hypospadias exhibit reduced anogenital distance. Hum Reprod 2012; 27:1577-80.

42. Sonke GS, Chang S, Strom SS, Sweeney AM, Annegers JF, Sigurdson AJ. Prenatal and perinatal risk factors and testicular cancer: a hospital-based case-control study. Oncol Res 2007; 16:383-7.

43. Weir HK, Marrett LD, Kreiger N, Darlington GA, Sugar L. Pre-natal and peri-natal exposures and risk of testicular germ-cell cancer. Int J Cancer 2000; 87:438-43.

44. McGlynn KA, Quraishi SM, Graubard BI, Weber JP, Rubertone MV, Erickson RL. Persistent organochlorine pesticides and risk of testicular germ cell tumors. J Natl Cancer Inst 2008; 100:663-71.

45. Anand P, Kunnumakkara AB, Sundaram C, Harikumar KB, Tharakan ST, Lai OS, et al. Cancer is a preventable disease that requires major lifestyle changes. Pharm Res 2008; 25:2097-116.

46. ibid.

47. Rahman I, Biswas SK, Kode A. Oxidant and antioxidant balance in the airways and airway diseases. Eur J Pharmacol 2006; 533:222-39.

48. Rahman I, Marwick J, Kirkham P. Redox modulation of chromatin remodeling: impact on histone acetylation and deacetylation, NF-kappaB and pro-inflammatory gene expression. Biochem Pharmacol 2004; 68:1255-67.

49. Joubert BR, Haberg SE, Nilsen RM, Wang X, Vollset SE, Murphy SK, et al. 450K epigenome-wide scan identifies differential DNA methylation in newborns related to maternal smoking during pregnancy. Environ Health Perspect 2012; 120:1425-31.

50. Stjernfeldt M, Berglund K, Lindsten J, Ludvigsson J. Maternal smoking during pregnancy and risk of childhood cancer. Lancet 1986; 1:1350-2.

51. Milne E, Greenop KR, Scott RJ, Bailey HD, Attia J, Dalla-Pozza L, et al. Parental prenatal smoking and risk of childhood acute lymphoblastic leukemia. Am J Epidemiol 2012; 175:43-53.

52. Chang JS, Selvin S, Metayer C, Crouse V, Golembesky A, Buffler PA. Parental smoking and the risk of childhood leukemia. Am J Epidemiol 2006; 163:1091-100.

53. Murugan S, Zhang C, Mojtahedzadeh S, Sarkar DK. Alcohol exposure in utero increases susceptibility to prostate tumorigenesis in rat offspring. Alcohol Clin Exp Res 2013.

54. MacArthur AC, McBride ML, Spinelli JJ, Tamaro S, Gallagher RP, Theriault G. Risk of childhood leukemia associated with parental smoking and alcohol consumption prior to conception and during pregnancy: the cross-Canada childhood leukemia study. Cancer Causes Control 2008; 19:283-95.

55. Doll R, Peto R. The causes of cancer: quantitative estimates of avoidable risks of cancer in the United States today. J Natl Cancer Inst 1981; 66:1191-308.

56. De Pergola G, Silvestris F. Obesity as a major risk factor for cancer. J Obes 2013; 2013:291546.

57. Calle EE, Rodriguez C, Walker-Thurmond K, Thun MJ. Overweight, obesity, and mortality from cancer in a prospectively studied cohort of U.S. adults. N Engl J Med 2003; 348:1625-38.

58. De Pergola G, Silvestris F. Obesity as a major risk factor for cancer. J Obes 2013; 2013:291546.

59. Cappellani A, Di Vita M, Zanghi A, Cavallaro A, Piccolo G, Veroux M, et al. Diet, obesity and breast cancer: an update. Front Biosci (Schol Ed) 2012; 4:90-108.

60. Wakai K, Matsuo K, Nagata C, Mizoue T, Tanaka K, Tsuji I, et al. Lung cancer risk and consumption of vegetables and fruit: an evaluation based on a systematic review of epidemiological evidence from Japan. Jpn J Clin Oncol 2011; 41:693-708.

61. Key TJ. Fruit and vegetables and cancer risk. Br J Cancer 2011; 104:6-11.

62. Lee CH, Ko AM, Warnakulasuriya S, Ling TY, Sunarjo, Rajapakse PS, et al. Population burden of betel quid abuse and its relation to oral premalignant disorders in South, Southeast, and East Asia: an Asian Betel-quid Consortium Study. Am J Public Health 2012; 102:e17-24.

63. Wu IC, Wu CC, Lu CY, Hsu WH, Wu MC, Lee JY, et al. Substance use (alcohol, areca nut and cigarette) is associated with poor prognosis of esophageal squamous cell carcinoma. PLoS One 2013; 8:e55834.

64. Liu Y, Wu F. Global burden of aflatoxin-induced hepatocellular carcinoma: a risk assessment. Environ Health Perspect 2010; 118:818-24.

65. Anand P, Kunnumakkara AB, Sundaram C, Harikumar KB, Tharakan ST, Lai OS, et al. Cancer is a preventable disease. Pharm Res 2008; 25:2097-116.

66. Kruk J, Czerniak U. Physical activity and its relation to cancer risk: updating the evidence. Asian Pac J Cancer Prev 2013; 14:3993-4003.

67. Colditz GA, Cannuscio CC, Frazier AL. Physical activity and reduced risk of colon cancer: implications for prevention. Cancer Causes Control 1997; 8:649-67.

68. Friedenreich CM. Physical activity and breast cancer: review of the epidemiologic evidence and biologic mechanisms. Recent Results Cancer Res 2011; 188:125-39.

69. Voskuil DW, Monninkhof EM, Elias SG, Vlems FA, van Leeuwen FE. Physical activity and endometrial cancer risk, a systematic review of current evidence. Cancer Epidemiol Biomarkers Prev 2007; 16:639-48.

70. Kruk J, Czerniak U. Physical activity and its relation to cancer risk: updating the evidence. Asian Pac J Cancer Prev 2013; 14:3993-4003.

71. Fasinu P, Orisakwe OE. Heavy metal pollution in sub-Saharan Africa and possible implications in cancer epidemiology. Asian Pac J Cancer Prev 2013; 14:3393-402.

72. Zhao Q, Wang Y, Cao Y, Chen A, Ren M, Ge Y, et al. Potential health risks of heavy metals in cultivated topsoil and grain,

including correlations with human primary liver, lung and gastric cancer, in Anhui province, Eastern China. Sci Total Environ 2013; 470-471C:340-7.

73. Parkin DM. The global health burden of infection-associated cancers in the year 2002. Int J Cancer 2006; 118:3030-44.

74. Maekita T, Nakazawa K, Mihara M, Nakajima T, Yanaoka K, Iguchi M, et al. High levels of aberrant DNA methylation in Helicobacter pylori-infected gastric mucosae and its possible association with gastric cancer risk. Clin Cancer Res 2006; 12:989-95.

75. Kitajima Y, Ohtaka K, Mitsuno M, Tanaka M, Sato S, Nakafusa Y, et al. Helicobacter pylori infection is an independent risk factor for Runx3 methylation in gastric cancer. Oncol Rep 2008; 19:197-202.

76. Um TH, Kim H, Oh BK, Kim MS, Kim KS, Jung G, et al. Aberrant CpG island hypermethylation in dysplastic nodules and early HCC of hepatitis B virus-related human multistep hepatocarcinogenesis. J Hepatol 2011; 54:939-47.

77. Wentzensen N, Sherman ME, Schiffman M, Wang SS. Utility of methylation markers in cervical cancer early detection: appraisal of the state-of-the-science. Gynecol Oncol 2009; 112:293-9.

78. Frazer I. Correlating immunity with protection for HPV infection. Int J Infect Dis 2007; 11 Suppl 2:S10-6.

79. Belpomme D, Irigaray P, Hardell L, Clapp R, Montagnier L, Epstein S, et al. The multitude and diversity of environmental carcinogens. Environ Res 2007; 105:414-29.

80. Ho SM, Johnson A, Tarapore P, Janakiram V, Zhang X, Leung YK. Environmental epigenetics and its implication. ILAR J 2012; 53:289-305

Chapter 10: Asthma, allergy and immune diseases

1. Prescott SL. The Allergy Epidemic: A Mystery of Modern Life. Perth: UWA Publishing; 2011.

2. Bostock J. On the cattarhus aestivus or summer catarrh. Medico-Chirurg Trans 1828; 14:437-46.

3. Hopper JL, Jenkins MA, Carlin JB, Giles GG. Increase in the self-reported prevalence of asthma and hay fever in adults over the last generation: a matched parent-offspring study. Aust J Public Health 1995; 19:120-4.

4. Nicholson JK, Holmes E, Kinross J, Burcelin R, Gibson G, Jia W, et al. Host-gut microbiota metabolic interactions. Science 2012; 336:1262-7.

5. Sommer F, Backhed F. The gut microbiota–masters of host development and physiology. Nat Rev Microbiol 2013; 11:227-38.

6. Mullins RJ. Paediatric food allergy trends in a community-based specialist allergy practice, 1995-2006. Med J Aust 2007; 186:618-21.

7. Osborne NJ, Koplin JJ, Martin PE, Gurrin LC, Lowe AJ, Matheson MC, et al. Prevalence of challenge-proven IgE-mediated food allergy using population-based sampling and predetermined challenge criteria in infants. J Allergy Clin Immunol 2011; 127:668-76 e1-2.

8. Palm NW, Rosenstein RK, Medzhitov R. Allergic host defences. Nature 2012; 484:465-72.

9. Bilbo SD, Drazen DL, Quan N, He L, Nelson RJ. Short day lengths attenuate the symptoms of infection in Siberian hamsters. Proc Biol Sci 2002; 269:447-54.

10. Yellon SM, Fagoaga OR, Nehlsen-Cannarella SL. Influence of photoperiod on immune cell functions in the male Siberian hamster. Am J Physiol 1999; 276:R97-R102.

11. Matarese G, La Cava A. The intricate interface between immune system and metabolism. Trends Immunol 2004; 25:193-200.

12. ibid.

13. Radon K, Schulze A, Schierl R, Dietrich-Gumperlein G, Nowak D, Jorres RA. Serum leptin and adiponectin levels and their association with allergic sensitization. Allergy 2008; 63:1448-54.

14. Matarese G, Leiter EH, La Cava A. Leptin in autoimmunity: many questions, some answers. Tissue Antigens 2007; 70:87-95.

15. Palmer G, Gabay C. A role for leptin in rheumatic diseases? Ann Rheum Dis 2003; 62:913-5.

16. Otsuka R, Yatsuya H, Tamakoshi K, Matsushita K, Wada K, Toyoshima H. Perceived psychological stress and serum leptin concentrations in Japanese men. Obesity (Silver Spring) 2006; 14:1832-8.

17. Hickey MS, Houmard JA, Considine RV, Tyndall GL, Midgette JB, Gavigan KE, et al. Gender-dependent effects of exercise training on serum leptin levels in humans. Am J Physiol 1997; 272:E562-6.

18. de Salles BF, Simao R, Fleck SJ, Dias I, Kraemer-Aguiar LG, Bouskela E. Effects of resistance training on cytokines. Int J Sports Med 2010; 31:441-50.

19. Visser M, Bouter LM, McQuillan GM, Wener MH, Harris TB. Elevated C-reactive protein levels in overweight and obese adults. JAMA 1999; 282:2131-5.

20. Visness CM, London SJ, Daniels JL, Kaufman JS, Yeatts KB, Siega-Riz AM, et al. Association of obesity with IgE levels and allergy

symptoms in children and adolescents: results from the National
Health and Nutrition Examination Survey 2005-2006. J Allergy Clin
Immunol 2009; 123:1163-9, 9 el-4.

21. Medzhitov R. Origin and physiological roles of inflammation.
Nature 2008; 454:428-35.

22. McDade TW. Early environments and the ecology of inflammation.
Proc Natl Acad Sci USA 2012; 109 Suppl 2:17281-8.

23. Gold DR, Damokosh AI, Dockery DW, Berkey CS. Body-mass
index as a predictor of incident asthma in a prospective cohort of
children. Pediatr Pulmonol 2003; 36:514-21.

24. Castro-Rodriguez JA, Holberg CJ, Morgan WJ, Wright AL,
Martinez FD. Increased incidence of asthmalike symptoms in girls
who become overweight or obese during the school years. Am J
Respir Crit Care Med 2001; 163:1344-9.

25. Ramsay JE, Ferrell WR, Crawford L, Wallace AM, Greer IA,
Sattar N. Maternal obesity is associated with dysregulation of
metabolic, vascular, and inflammatory pathways. J Clin Endocrinol
Metab 2002; 87:4231-7.

26. Challier JC, Basu S, Bintein T, Minium J, Hotmire K, Catalano PM,
et al. Obesity in pregnancy stimulates macrophage accumulation and
inflammation in the placenta. Placenta 2008; 29:274-81.

27. Reichman NE, Nepomnyaschy L. Maternal pre-pregnancy obesity
and diagnosis of asthma in offspring at age 3 years. Matern Child
Health J 2008; 12:725-33.

28. Haberg SE, Stigum H, London SJ, Nystad W, Nafstad P. Maternal
obesity in pregnancy and respiratory health in early childhood.
Paediatr Perinat Epidemiol 2009; 23:352-62.

29. Rothenbacher D, Weyermann M, Fantuzzi G, Brenner H.
Adipokines in cord blood and risk of wheezing disorders within the
first two years of life. Clin Exp Allergy 2007; 37:1143-9.

30. Radon K, Schulze A, Schierl R, Dietrich-Gumperlein G, Nowak D,
Jorres RA. Serum leptin and adiponectin levels. Allergy 2008;
63:1448-54.

31. Ciprandi G, De Amici M, Tosca MA, Marseglia G. Serum leptin
levels depend on allergen exposure in patients with seasonal allergic
rhinitis. Immunol Invest 2009; 38:681-9. 32.   Matarese G, La
Cava A. The intricate interface between immune system and
metabolism. Trends Immunol 2004; 25:193-200.

33. Knoflach M, Kiechl S, Mayr A, Willeit J, Poewe W, Wick G.
Allergic rhinitis, asthma, and atherosclerosis in the Bruneck and
ARMY studies. Arch Intern Med 2005; 165:2521-6.

34. Onufrak S, Abramson J, Vaccarino V. Adult-onset asthma is associated with increased carotid atherosclerosis among women in the Atherosclerosis Risk in Communities (ARIC) study. Atherosclerosis 2007; 195:129-37.

35. Matheson EM, Player MS, Mainous AG, 3rd, King DE, Everett CJ. The association between hay fever and stroke in a cohort of middle aged and elderly adults. J Am Board Fam Med 2008; 21:179-83.

36. Iribarren C, Tolstykh IV, Miller MK, Sobel E, Eisner MD. Adult asthma and risk of coronary heart disease, cerebrovascular disease, and heart failure: a prospective study of 2 matched cohorts. Am J Epidemiol 2012; 176:1014-24.

37. Knoflach M, Kiechl S, Mayr A, Willeit J, Poewe W, Wick G. Allergic rhinitis, asthma, and atherosclerosis. Arch Intern Med 2005; 165:2521-6.

38. Brunekreef B, Hoek G, Fischer P, Spieksma FT. Relation between airborne pollen concentrations and daily cardiovascular and respiratory-disease mortality. Lancet 2000; 355:1517-8.

39. Bilbo SD, Schwarz JM. The immune system and developmental programming of brain and behavior. Front Neuroendocrinol 2012; 33:267-86.

40. Pandey GN, Rizavi HS, Ren X, Fareed J, Hoppensteadt DA, Roberts RC, et al. Proinflammatory cytokines in the prefrontal cortex of teenage suicide victims. J Psychiatr Res 2012; 46:57-63.

41. Pandey GN, Rizavi HS, Ren X, Dwivedi Y, Palkovits M. Region-specific alterations in glucocorticoid receptor expression in the postmortem brain of teenage suicide victims. Psychoneuroendocrinology 2013.

42. Postolache TT, Mortensen PB, Tonelli LH, Jiao X, Frangakis C, Soriano JJ, et al. Seasonal spring peaks of suicide in victims with and without prior history of hospitalization for mood disorders. J Affect Disord 2010; 121:88-93.

43. Tonelli LH, Stiller J, Rujescu D, Giegling I, Schneider B, Maurer K, et al. Elevated cytokine expression in the orbitofrontal cortex of victims of suicide. Acta Psychiatr Scand 2008; 117:198-206.

44. Tonelli LH, Hoshino A, Katz M, Postolache TT. Acute stress promotes aggressive-like behavior in rats made allergic to tree pollen. Int J Child Health Hum Dev 2008; 1:305-12.

45. Tonelli LH, Katz M, Kovacsics CE, Gould TD, Joppy B, Hoshino A, et al. Allergic rhinitis induces anxiety-like behavior and altered social interaction in rodents. Brain Behav Immun 2009; 23:784-93.

46. Stenius F, Borres M, Bottai M, Lilja G, Lindblad F, Pershagen G, et al. Salivary cortisol levels and allergy in children: The ALADDIN birth cohort. J Allergy Clin Immunol 2011; 28:1335-39.
47. Chida Y, Hamer M, Steptoe A. A bidirectional relationship between psychosocial factors and atopic disorders: a systematic review and meta-analysis. Psychosom Med 2008; 70:102-16.
48. Mold JE, McCune JM. At the crossroads between tolerance and aggression: Revisiting the 'layered immune system' hypothesis. Chimerism 2011; 2:35-41.
49. Prescott S, Saffery R. The role of epigenetic dysregulation in the epidemic of allergic disease. Clin Epigenetics 2011; 2:223-32.
50. Prescott S, Macaubas C, Smallacombe T, Holt B, Sly P, Loh R, et al. Development of allergen-specific T-cell memory in atopic and normal children. Lancet 1999; 353(9148):196-200.
51. Wegmann TG, Lin H, Guilbert L, Mosmann TR. Bidirectional cytokine interactions in the maternal-fetal relationship: is successful pregnancy a Th2 phenomenon? Immunol Today 1993; 14 (7):353-56.
52. Prescott S, Macaubas C, Smallacombe T, Holt B, Sly P, Loh R, et al. Development of allergen-specific T-cell memory. Lancet 1999; 353(9148):196-200.
53. Schaub B, Liu J, Hoppler S, Schleich I, Huehn J, Olek S, et al. Maternal farm exposure modulates neonatal immune mechanisms through regulatory T cells. J Allergy Clin Immunol 2009; 123:774-82 e5.
54. West CE, Jenmalm MC, Prescott SL. The gut microbiota and its role in the development of allergic disease: a wider perspective. Clin Exp Allergy 2014.
55. Brand S, Teich R, Dicke T, Harb H, Yildirim AO, Tost J, et al. Epigenetic regulation in murine offspring as a novel mechanism for transmaternal asthma protection induced by microbes. J Allergy Clin Immunol 2011; 128:618-25 e1-7.
56. Ege MJ, Bieli C, Frei R, van Strien RT, Riedler J, Ublagger E, et al. Prenatal farm exposure is related to the expression of receptors of the innate immunity and to atopic sensitization in school-age children. J Allergy Clin Immunol 2006; 117:817-23.
57. Douwes J, Cheng S, Travier N, Cohet C, Niesink A, McKenzie J, et al. Farm exposure in utero may protect against asthma, hay fever and eczema. Eur Respir J 2008; 32:603-11.
58. Riedler J, Eder W, Oberfeld G, Schreuer M. Austrian children living on a farm have less hay fever, asthma and allergic sensitization. Clin Exp Allergy 2000; 30:194-200.

59. Ege MJ, Bieli C, Frei R, van Strien RT, Riedler J, Ublagger E, et al. Prenatal farm exposure is related to the expression of receptors. J Allergy Clin Immunol 2006; 117:817-23.

60. Pfefferle PI, Prescott SL, Kopp M. Microbial influence on tolerance and opportunities for intervention with prebiotics/probiotics and bacterial lysates. J Allergy Clin Immunol 2013; 131:1453-63; quiz 64.

61. Pelucchi C, Chatenoud L, Turati F, Galeone C, Moja L, Bach JF, et al. Probiotics supplementation during pregnancy or infancy for the prevention of atopic dermatitis: a meta-analysis. Epidemiology 2012; 23:402-14.

62. Bertelsen RJ, Brantsaeter AL, Magnus MC, Haugen M, Myhre R, Jacobsson B, et al. Probiotic milk consumption in pregnancy and infancy and subsequent childhood allergic diseases. J Allergy Clin Immunol 2013.

63. Dunstan J, Mori TA, Barden A, Beilin LJ, Taylor A, Holt PG, et al. Fish oil supplementation in pregnancy modifies neonatal allergen-specific immune responses and clinical outcomes in infants at high risk of atopy: a randomised controlled trial. J Allergy Clin Immunol 2003; 112:1178-84.

64. Kiefte-de Jong JC, Timmermans S, Jaddoe VW, Hofman A, Tiemeier H, Steegers EA, et al. High circulating folate and vitamin B-12 concentrations in women during pregnancy are associated with increased prevalence of atopic dermatitis in their offspring. J Nutr 2012; 142:731-8.

65. Dunstan JA, West C, McCarthy S, Metcalfe J, Meldrum S, Oddy WH, et al. The relationship between maternal folate status in pregnancy, cord blood folate levels, and allergic outcomes in early childhood. Allergy 2012; 67:50-7.

66. Håberg SE, London SJ, Stigum H, Nafstad P, Nystad W. Folic acid supplements in pregnancy and early childhood respiratory health. Arch Dis Child 2009; 94:180-4.

67. Whitrow MJ, Moore VM, Rumbold AR, Davies MJ. Effect of supplemental folic acid in pregnancy on childhood asthma: a prospective birth cohort study. Am J Epidemiol 2009; 170:1486-93.

68. Liu J, Ballaney M, Al-alem U, Quan C, Jin X, Perera F, et al. Combined inhaled diesel exhaust particles and allergen exposure alter methylation of T helper genes and IgE production in vivo. Toxicol Sci 2008; 102:76-81.

69. Perera F, Tang WY, Herbstman J, Tang D, Levin L, Miller R, et al. Relation of DNA methylation of 5'-CpG island of ACSL3 to transplacental exposure to airborne polycyclic aromatic hydrocarbons and childhood asthma. PLoS One 2009; 4:e4488.

70. Daniel V, Huber W, Bauer K, Opelz G. Impaired in-vitro lymphocyte responses in patients with elevated pentachlorophenol (PCP) blood levels. Arch. Environ. Health 1995; 50 (4): 287-92.

71. Noakes PS, Taylor P, Wilkinson S, Prescott SL. The relationship between persistent organic pollutants in maternal and neonatal tissues and immune responses to allergens: A novel exploratory study. Chemosphere 2006; 63:1304-11.

72. Shaheen SO, Newson RB, Sherriff A, Henderson AJ, Heron JE, Burney PG, et al. Paracetamol use in pregnancy and wheezing in early childhood. Thorax 2002; 57:958-63.

73. Rebordosa C, Kogevinas M, Sorensen HT, Olsen J. Pre-natal exposure to paracetamol and risk of wheezing and asthma in children: a birth cohort study. Int J Epidemiol 2008; 37:583-90.

74. Bisgaard H, Loland L, Holst KK, Pipper CB. Prenatal determinants of neonatal lung function in high-risk newborns. J Allergy Clin Immunol 2009; 123:651-7, 7 e1-4.

75. Garcia-Marcos L, Sanchez-Solis M, Perez-Fernandez V, Pastor-Vivero MD, Mondejar-Lopez P, Valverde-Molina J. Is the Effect of Prenatal Paracetamol Exposure on Wheezing in Preschool Children Modified by Asthma in the Mother? Int Arch Allergy Immunol 2008; 149:33-7.

76. Persky V, Piorkowski J, Hernandez E, Chavez N, Wagner-Cassanova C, Vergara C, et al. Prenatal exposure to acetaminophen and respiratory symptoms in the first year of life. Ann Allergy Asthma Immunol 2008; 101:271-8.

77. Dehlink E, Yen E, Leichtner AM, Hait EJ, Fiebiger E. First evidence of a possible association between gastric acid suppression during pregnancy and childhood asthma: a population-based register study. Clin Exp Allergy 2009; 39:246-53.

78. Thavagnanam S, Fleming J, Bromley A, Shields MD, Cardwell CR. A meta-analysis of the association between Caesarean section and childhood asthma. Clin Exp Allergy 2008; 38:629-33.

79. Bager P, Wohlfahrt J, Westergaard T. Caesarean delivery and risk of atopy and allergic disease: meta-analyses. Clin Exp Allergy 2008; 38:634-42.

80. Tollanes MC, Moster D, Daltveit AK, Irgens LM. Cesarean section and risk of severe childhood asthma: a population-based cohort study. J Pediatr 2008; 153:112-6.

81. Metsala J, Kilkkinen A, Kaila M, Tapanainen H, Klaukka T, Gissler M, et al. Perinatal factors and the risk of asthma in childhood–a population-based register study in Finland. Am J Epidemiol 2008; 168:170-8.

82. Pistiner M, Gold DR, Abdulkerim H, Hoffman E, Celedon JC. Birth by cesarean section, allergic rhinitis, and allergic sensitization among children with a parental history of atopy. J Allergy Clin Immunol 2008; 122:274-9.

83. Jakobsson HE, Abrahamsson TR, Jenmalm MC, Harris K, Quince C, Jernberg C, et al. Decreased gut microbiota diversity, delayed Bacteroidetes colonisation and reduced Th1 responses in infants delivered by Caesarean section. Gut 2013.

84. Kuitunen M, Kukkonen K, Juntunen-Backman K, Korpela R, Poussa T, Tuure T, et al. Probiotics prevent IgE-associated allergy until age 5 years in cesarean-delivered children but not in the total cohort. J Allergy Clin Immunol 2009; 123:335-41.

85. Kallen B, Finnstrom O, Nygren KG, Otterblad Olausson P. Asthma in Swedish children conceived by in vitro fertilisation. Arch Dis Child 2013; 98:92-6.

86. Fox AT, Sasieni P, du Toit G, Syed H, Lack G. Household peanut consumption as a risk factor for the development of peanut allergy. J Allergy Clin Immunol 2009; 123:417-23.

87. Venter C, Pereira B, Voigt K, Grundy J, Clayton CB, Higgins B, et al. Factors associated with maternal dietary intake, feeding and weaning practices, and the development of food hypersensitivity in the infant. Pediatr Allergy Immunol 2009.

88. Willers SM, Wijga AH, Brunekreef B, Kerkhof M, Gerritsen J, Hoekstra MO, et al. Maternal food consumption during pregnancy and the longitudinal development of childhood asthma. Am J Respir Crit Care Med 2008; 178:124-31.

89. Metcalfe J, Prescott SL, Palmer DJ. Randomized controlled trials investigating the role of allergen exposure in food allergy: where are we now? Curr Opin Allergy Clin Immunol 2013; 13:296-305.

90. Sudo N, Sawamura S, Tanaka K, Aiba Y, Kubo C, Koga Y. The requirement of intestinal bacterial flora for the development of an IgE production system fully susceptible to oral tolerance induction. J Immunol 1997; 159:1739-45.

91. Prescott SL. Early-life environmental determinants of allergic diseases and the wider pandemic of inflammatory noncommunicable diseases. J Allergy Clin Immunol 2013; 131:23-30.

92 Moro G, Arslanoglu S, Stahl B, Jelinek J, Wahn U, Boehm G. A mixture of prebiotic oligosaccharides reduces the incidence of atopic dermatitis during the first six months of age. Arch Dis Child 2006; 91:814-9.

93. Boehm G, Lidestri M, Casetta P, Jelinek J, Negretti F, Stahl B, et al. Supplementation of a bovine milk formula with an oligosaccharide

mixture increases counts of faecal bifidobacteria in preterm infants. Arch Dis Child Fetal Neonatal Ed 2002; 86:F178-81.

94. Bouhnik Y, Vahedi K, Achour L, Attar A, Salfati J, Pochart P, et al. Short-chain fructo-oligosaccharide administration dose-dependently increases fecal bifidobacteria in healthy humans. J Nutr 1999; 129:113-6.

95. Maslowski KM, Vieira AT, Ng A, Kranich J, Sierro F, Yu D, et al. Regulation of inflammatory responses by gut microbiota and chemoattractant receptor GPR43. Nature 2009; 461:1282-6.

96. Vulevic J, Juric A, Tzortzis G, Gibson GR. A mixture of trans-galactooligosaccharides reduces markers of metabolic syndrome and modulates the fecal microbiota and immune function of overweight adults. J Nutr 2013; 143(3):324-31.

97. Arslanoglu S, Moro GE, Schmitt J, Tandoi L, Rizzardi S, Boehm G. Early dietary intervention with a mixture of prebiotic oligosaccharides reduces the incidence of allergic manifestations and infections during the first two years of life. J Nutr 2008; 138(6):1091-5.

98. Gruber C, van Stuijvenberg M, Mosca F, Moro G, Chirico G, Braegger CP, et al. Reduced occurrence of early atopic dermatitis because of immunoactive prebiotics among low-atopy-risk infants. J Allergy Clin Immunol 2010; ePub ahead of print.

99. Pfefferle PI, Prescott SL, Kopp M. Microbial influence on tolerance. J Allergy Clin Immunol 2013; 131:1453-63; quiz 64.

100. Marshall BJ, Warren JR. Unidentified curved bacilli on gastric epithelium in active chronic gastritis. Lancet 1983; 1:1273-5.

101. Marshall BJ, Warren JR. Unidentified curved bacilli in the stomach of patients with gastritis and peptic ulceration. Lancet 1984; 1:1311-5.

102. Amedei A, Codolo G, Del Prete G, de Bernard M, D'Elios MM. The effect of Helicobacter pylori on asthma and allergy. J Asthma Allergy 2010; 3:139-47.

103. Kosunen TU, Hook-Nikanne J, Salomaa A, Sarna S, Aromaa A, Haahtela T. Increase of allergen-specific immunoglobulin E antibodies from 1973 to 1994 in a Finnish population and a possible relationship to Helicobacter pylori infections. Clin Exp Allergy 2002; 32:373-8.

104. Chen Y, Blaser MJ. Inverse associations of Helicobacter pylori with asthma and allergy. Arch Intern Med 2007; 167:821-7.

105. Guarner F, Bourdet-Sicard R, Brandtzaeg P, Gill HS, McGuirk P, van Eden W, et al. Mechanisms of disease: the hygiene hypothesis revisited. Nat Clin Pract Gastroenterol Hepatol 2006; 3:275-84.

106. Arnold IC, Hitzler I, Muller A. The immunomodulatory properties of Helicobacter pylori confer protection against allergic and chronic inflammatory disorders. Front Cell Infect Microbiol 2012; 2:10.

107. Arnold IC, Dehzad N, Reuter S, Martin H, Becher B, Taube C, et al. Helicobacter pylori infection prevents allergic asthma in mouse models through the induction of regulatory T cells. J Clin Invest 2011; 121:3088-93.

108. Prescott SL, Smith P, Tang MLK, Palmer DJ, Sinn J, Huntley SJ, et al. The importance of early complementary feeding in the development of oral tolerance: concerns and controversies. Pediatr Allergy Immunol 2008; 19(5) 375-80.

109. Greer FR, Sicherer SH, Burks AW. Effects of early nutritional interventions on the development of atopic disease in infants and children: the role of maternal dietary restriction, breastfeeding, timing of introduction of complementary foods, and hydrolyzed formulas. Pediatrics 2008; 121:183-91.

110. Host A, Halken S, Muraro A, Dreborg S, Niggemann B, Aalberse R, et al. Dietary prevention of allergic diseases in infants and small children. Pediatr Allergy Immunol 2008; 19:1-4.

111. Agostoni C, Decsi T, Fewtrell M, Goulet O, Kolacek S, Koletzko B, et al. Complementary feeding: a commentary by the ESPGHAN Committee on Nutrition. J Pediatr Gastroenterol Nutr 2008; 46:99-110.

112. Palmer DJ, Metcalfe J, Makrides M, Gold MS, Quinn P, West CE, et al. Early regular egg exposure in infants with eczema: A randomized controlled trial. J Allergy Clin Immunol 2013; 132:387-92 e1.

Chapter 11: Healthy ageing starts in early life: genetics, epigenetics and telomeres

1. Shakespeare W. As You Like It, Act 2, Scene 7.

2. Baker DJ, Wijshake T, Tchkonia T, LeBrasseur NK, Childs BG, van de Sluis B, et al. Clearance of p16Ink4a-positive senescent cells delays ageing-associated disorders. Nature 2011; 479:232-6.

3. Pyo JO, Yoo SM, Jung YK. The Interplay between Autophagy and Aging. Diabetes Metab J 2013; 37:333-9.

4. Johnson SC, Rabinovitch PS, Kaeberlein M. mTOR is a key modulator of ageing and age-related disease. Nature 2013; 493:338-45.

5. Speakman JR, Mitchell SE. Caloric restriction. Mol Aspects Med 2011; 32:159-221.

6. Rowe JW, Kahn RL. Successful aging. Gerontologist 1997;
    37:433-40.
7. Barja G. Free radicals and aging. Trends Neurosci 2004; 27:595-600.
8. Freitas AA, de Magalhaes JP. A review and appraisal of the DNA
    damage theory of ageing. Mutat Res 2011; 728:12-22.
9. Hayflick L, Moorhead PS. The serial cultivation of human diploid
    cell strains. Exp Cell Res 1961; 25:585-621.
10. Shay JW, Wright WE. Hayflick, his limit, and cellular ageing. Nat
    Rev Mol Cell Biol 2000; 1:72-6.
11. Hayflick L. The cell biology of human aging. N Engl J Med 1976;
    295:1302-8.
12. Hayflick L. The cell biology of human aging. N Engl J Med 1976;
    295:1302-8.
13. Blackburn EH, Gall JG. A tandemly repeated sequence at the termini
    of the extrachromosomal ribosomal RNA genes in Tetrahymena.
    J Mol Biol 1978; 120:33-53.
14. Bojesen SE. Telomeres and human health. J Intern Med 2013;
    274:399-413.
15. Mather KA, Jorm AF, Parslow RA, Christensen H. Is telomere
    length a biomarker of aging? A review. J Gerontol A Biol Sci Med Sci
    2011; 66:202-13.
16. Shalev I, Entringer S, Wadhwa PD, Wolkowitz OM, Puterman E,
    Lin J, et al. Stress and telomere biology: a lifespan perspective.
    Psychoneuroendocrinology 2013; 38:1835-42.
17. Hallows SE, Regnault TR, Betts DH. The long and short of it: the
    role of telomeres in fetal origins of adult disease. J Pregnancy 2012;
    2012:638476.
18. Kim Sh SH, Kaminker P, Campisi J. Telomeres, aging and cancer: in
    search of a happy ending. Oncogene 2002; 21:503-11.
19. Olovnikov AM. A theory of marginotomy. The incomplete copying
    of template margin in enzymic synthesis of polynucleotides and
    biological significance of the phenomenon. J Theor Biol 1973;
    41:181-90.
20. Greider CW, Blackburn EH. Identification of a specific telomere
    terminal transferase activity in Tetrahymena extracts. Cell 1985;
    43:405-13.
21. Bodnar AG, Ouellette M, Frolkis M, Holt SE, Chiu CP, Morin GB,
    et al. Extension of life-span by introduction of telomerase into
    normal human cells. Science 1998; 279:349-52.
22. Jaskelioff M, Muller FL, Paik JH, Thomas E, Jiang S, Adams AC,
    et al. Telomerase reactivation reverses tissue degeneration in aged
    telomerase-deficient mice. Nature 2011; 469:102-6.

23. Callaway E. Telomerase reverses ageing process. Nature News, 2010.
24. Bernardes de Jesus B, Vera E, Schneeberger K, Tejera AM, Ayuso E, Bosch F, et al. Telomerase gene therapy in adult and old mice delays aging and increases longevity without increasing cancer. EMBO Mol Med 2012; 4:691-704.
25. Hastie ND, Dempster M, Dunlop MG, Thompson AM, Green DK, Allshire RC. Telomere reduction in human colorectal carcinoma and with ageing. Nature 1990; 346:866-8.
26. Gardner M, Bann D, Wiley L, Cooper R, Hardy R, Nitsch D, et al. Gender and telomere length: Systematic review and meta-analysis. Exp Gerontol 2013.
27. Okuda K, Bardeguez A, Gardner JP, Rodriguez P, Ganesh V, Kimura M, et al. Telomere length in the newborn. Pediatr Res 2002; 52:377-81.
28. Butt HZ, Atturu G, London NJ, Sayers RD, Bown MJ. Telomere length dynamics in vascular disease: a review. Eur J Vasc Endovasc Surg 2010; 40:17-26.
29. ibid.
30. Albrecht E, Sillanpaa E, Karrasch S, Alves AC, Codd V, Hovatta I, et al. Telomere length in circulating leukocytes is associated with lung function and disease. Eur Respir J 2013.
31. Rode L, Bojesen SE, Weischer M, Vestbo J, Nordestgaard BG. Short telomere length, lung function and chronic obstructive pulmonary disease in 46,396 individuals. Thorax 2013; 68:429-35.
32. Tzanetakou IP, Katsilambros NL, Benetos A, Mikhailidis DP, Perrea DN. "Is obesity linked to aging?": adipose tissue and the role of telomeres. Ageing Res Rev 2012; 11:220-9.
33. Ma D, Zhu W, Hu S, Yu X, Yang Y. Association between oxidative stress and telomere length in type 1 and type 2 diabetic patients. J Endocrinol Invest 2013.
34. Valdes AM, Deary IJ, Gardner J, Kimura M, Lu X, Spector TD, et al. Leukocyte telomere length is associated with cognitive performance in healthy women. Neurobiol Aging 2010; 31:986-92.
35. Ma H, Zhou Z, Wei S, Liu Z, Pooley KA, Dunning AM, et al. Shortened telomere length is associated with increased risk of cancer: a meta-analysis. PLoS One 2011; 6:e20466.
36. Wentzensen IM, Mirabello L, Pfeiffer RM, Savage SA. The association of telomere length and cancer: a meta-analysis. Cancer Epidemiol Biomarkers Prev 2011; 20:1238-50.
37. Weischer M, Nordestgaard BG, Cawthon RM, Freiberg JJ, Tybjaerg-Hansen A, Bojesen SE. Short telomere length, cancer survival, and cancer risk in 47102 individuals. J Natl Cancer Inst 2013; 105:459-68.

38. Bojesen SE. Telomeres and human health. J Intern Med 2013; 274:399-413.
39. Mather KA, Jorm AF, Parslow RA, Christensen H. Is telomere length a biomarker of aging? A review. J Gerontol A Biol Sci Med Sci 2011; 66:202-13.
40. Blackburn E. Telomeres and Tetrahymena: an interview with Elizabeth Blackburn. Dis Model Mech 2009; 2:534-7.
41. Valdes AM, Andrew T, Gardner JP, Kimura M, Oelsner E, Cherkas LF, et al. Obesity, cigarette smoking, and telomere length in women. Lancet 2005; 366:662-4.
42. ibid.
43. Njajou OT, Cawthon RM, Blackburn EH, Harris TB, Li R, Sanders JL, et al. Shorter telomeres are associated with obesity and weight gain in the elderly. Int J Obes (Lond) 2012; 36:1176-9.
44. Cherkas LF, Hunkin JL, Kato BS, Richards JB, Gardner JP, Surdulescu GL, et al. The association between physical activity in leisure time and leukocyte telomere length. Arch Intern Med 2008; 168:154-8.
45. Jackowska M, Hamer M, Carvalho LA, Erusalimsky JD, Butcher L, Steptoe A. Short sleep duration is associated with shorter telomere length in healthy men: findings from the Whitehall II cohort study. PLoS One 2012; 7:e47292.
46. Liang G, Schernhammer E, Qi L, Gao X, De Vivo I, Han J. Associations between rotating night shifts, sleep duration, and telomere length in women. PLoS One 2011; 6:e23462.
47. Prather AA, Puterman E, Lin J, O'Donovan A, Krauss J, Tomiyama AJ, et al. Shorter leukocyte telomere length in midlife women with poor sleep quality. J Aging Res 2011; 2011:721390.
48. Wolkowitz OM, Mellon SH, Epel ES, Lin J, Dhabhar FS, Su Y, et al. Leukocyte telomere length in major depression: correlations with chronicity, inflammation and oxidative stress—preliminary findings. PLoS One 2011; 6:e17837.
49. O'Donovan A, Pantell MS, Puterman E, Dhabhar FS, Blackburn EH, Yaffe K, et al. Cumulative inflammatory load is associated with short leukocyte telomere length in the Health, Aging and Body Composition Study. PLoS One 2011; 6:e19687.
50. Collerton J, Martin-Ruiz C, Davies K, Hilkens CM, Isaacs J, Kolenda C, et al. Frailty and the role of inflammation, immunosenescence and cellular ageing in the very old: cross-sectional findings from the Newcastle 85+ Study. Mech Ageing Dev 2012; 133:456-66.

51. O'Donovan A, Pantell MS, Puterman E, Dhabhar FS, Blackburn EH, Yaffe K, et al. Cumulative inflammatory load is associated with short leukocyte telomere length. PLoS One 2011; 6:e19687.

52. Hallows SE, Regnault TR, Betts DH. The long and short of it. J Pregnancy 2012; 2012:638476.

53. Entringer S, Epel ES, Lin J, Buss C, Shahbaba B, Blackburn EH, et al. Maternal psychosocial stress during pregnancy is associated with newborn leukocyte telomere length. Am J Obstet Gynecol 2013; 208:134 e1-7.

54. Shalev I, Moffitt TE, Sugden K, Williams B, Houts RM, Danese A, et al. Exposure to violence during childhood is associated with telomere erosion from 5 to 10 years of age: a longitudinal study. Mol Psychiatry 2013; 18:576-81.

55. Epel ES, Blackburn EH, Lin J, Dhabhar FS, Adler NE, Morrow JD, et al. Accelerated telomere shortening in response to life stress. Proc Natl Acad Sci U S A 2004; 101:17312-5.

56. Wolkowitz OM, Mellon SH, Epel ES, Lin J, Dhabhar FS, Su Y, et al. Leukocyte telomere length in major depression. PLoS One 2011; 6:e17837.

57. Brydon L, Lin J, Butcher L, Hamer M, Erusalimsky JD, Blackburn EH, et al. Hostility and cellular aging in men from the Whitehall II cohort. Biol Psychiatry 2012; 71:767-73.

58. Lee DC, Im JA, Kim JH, Lee HR, Shim JY. Effect of long-term hormone therapy on telomere length in postmenopausal women. Yonsei Med J 2005; 46:471-9.

59. Alkan C, Kidd JM, Marques-Bonet T, Aksay G, Antonacci F, Hormozdiari F, et al. Personalized copy number and segmental duplication maps using next-generation sequencing. Nat Genet 2009; 41:1061-7.

60. Iafrate AJ, Feuk L, Rivera MN, Listewnik ML, Donahoe PK, Qi Y, et al. Detection of large-scale variation in the human genome. Nat Genet 2004; 36:949-51.

61. Njajou OT, Cawthon RM, Damcott CM, Wu SH, Ott S, Garant MJ, et al. Telomere length is paternally inherited and is associated with parental lifespan. Proc Natl Acad Sci U S A 2007; 104:12135-9.

62. Slagboom PE, Droog S, Boomsma DI. Genetic determination of telomere size in humans: a twin study of three age groups. Am J Hum Genet 1994; 55:876-82.

63. Le Souef PN, Goldblatt J, Lynch NR. Evolutionary adaptation of inflammatory immune responses in human beings. Lancet 2000; 356:242-4.

64. Atzmon G, Cho M, Cawthon RM, Budagov T, Katz M, Yang X, et al. Evolution in health and medicine Sackler colloquium: Genetic variation in human telomerase is associated with telomere length in Ashkenazi centenarians. Proc Natl Acad Sci U S A 2010; 107 Suppl 1:1710-7.

65. Codd V, Nelson CP, Albrecht E, Mangino M, Deelen J, Buxton JL, et al. Identification of seven loci affecting mean telomere length and their association with disease. Nat Genet 2013; 45:422-7, 7e1-2.

66. ibid.

67. Brooks-Wilson AR. Genetics of healthy aging and longevity. Hum Genet 2013; 132:1323-38.

68. Beekman M, Blanche H, Perola M, Hervonen A, Bezrukov V, Sikora E, et al. Genome-wide linkage analysis for human longevity: Genetics of Healthy Aging Study. Aging Cell 2013; 12:184-93.

69. Di Bona D, Accardi G, Virruso C, Candore G, Caruso C. Association Between Genetic Variations In The Insulin/Insulin-Like Growth Factor (Igf-1) Signaling Pathway And Longevity: A Systematic Review And Meta-Analysis. Curr Vasc Pharmacol 2013.

70. Yang L, Froberg JE, Lee JT. Long noncoding RNAs: fresh perspectives into the RNA world. Trends Bio Sci 2014; 39:35-43.

71. Waterland RA, Michels KB. Epigenetic epidemiology of the developmental origins hypothesis. Annu Rev Nutr 2007; 27:363-88.

72. Brooks-Wilson AR. Genetics of healthy aging and longevity. Hum Genet 2013; 132:1323-38.

73. Fraga MF, Ballestar E, Paz MF, Ropero S, Setien F, Ballestar ML, et al. Epigenetic differences arise during the lifetime of monozygotic twins. Proc Natl Acad Sci U S A 2005; 102:10604-9.

74. Talens RP, Christensen K, Putter H, Willemsen G, Christiansen L, Kremer D, et al. Epigenetic variation during the adult lifespan: cross-sectional and longitudinal data on monozygotic twin pairs. Aging Cell 2012; 11:694-703.

75. Mill J, Dempster E, Caspi A, Williams B, Moffitt T, Craig I. Evidence for monozygotic twin (MZ) discordance in methylation level at two CpG sites in the promoter region of the catechol-O-methyltransferase (COMT) gene. Am J Med Genet B Neuropsychiatr Genet 2006; 141B:421-5.

76. Martino DJ, Tulic MK, Gordon L, Hodder M, Richman T, Metcalfe J, et al. Evidence for age-related and individual-specific changes in DNA methylation profile of mononuclear cells during early immune development in humans. Epigenetics 2011; 6.

77. Gravina S, Vijg J. Epigenetic factors in aging and longevity. Pflugers Arch 2010; 459:247-58

78. Ornish D, Lin J, Chan JM, Epel E, Kemp C, Weidner G, et al. Effect of comprehensive lifestyle changes on telomerase activity and telomere length in men with biopsy-proven low-risk prostate cancer: 5-year follow-up of a descriptive pilot study. Lancet Oncol 2013; 14:1112-20.

79. Ornish D, Lin J, Daubenmier J, Weidner G, Epel E, Kemp C, et al. Increased telomerase activity and comprehensive lifestyle changes: a pilot study. Lancet Oncol 2008; 9:1048-57.

80. Fontana L, Meyer TE, Klein S, Holloszy JO. Long-term calorie restriction. Proc Natl Acad Sci U S A 2004; 101:6659-63.

81. Garcia-Calzon S, Gea A, Razquin C, Corella D, Lamuela-Raventos RM, Martinez JA, et al. Longitudinal association of telomere length and obesity indices in an intervention study with a Mediterranean diet: the PREDIMED-NAVARRA trial. Int J Obes (Lond) 2013.

82. McCay CCM, Maynard LA. The effect of retarded growth upon the length of the lifespan and upon the ultimate body size. J Nutr 1935; 10:63-79.

83. Speakman JR, Mitchell SE. Caloric restriction. Mol Aspects Med 2011; 32:159-221.

84. Vera E, Bernardes de Jesus B, Foronda M, Flores JM, Blasco MA. Telomerase reverse transcriptase synergizes with calorie restriction to increase health span and extend mouse longevity. PLoS One 2013; 8:e53760.

85. Speakman JR, Mitchell SE. Caloric restriction. Mol Aspects Med 2011; 32:159-221.

86. Mattson MP. Energy intake, meal frequency, and health: a neurobiological perspective. Annu Rev Nutr 2005; 25:237-60.

87. Araya AV, Orellana X, Espinoza J. Evaluation of the effect of caloric restriction on serum BDNF in overweight and obese subjects: preliminary evidences. Endocrine 2008; 33:300-4.

88. Mattison JA, Roth GS, Beasley TM, Tilmont EM, Handy AM, Herbert RL, et al. Impact of caloric restriction on health and survival in rhesus monkeys from the NIA study. Nature 2012; 489:318-21.

89. Vera E, Bernardes de Jesus B, Foronda M, Flores JM, Blasco MA. Telomerase reverse transcriptase synergizes with calorie restriction. PLoS One 2013; 8:e53760.

90. Speakman JR, Mitchell SE. Caloric restriction. Mol Aspects Med 2011; 32:159-221.

91. Harrison DE, Strong R, Sharp ZD, Nelson JF, Astle CM, Flurkey K, et al. Rapamycin fed late in life extends lifespan in genetically heterogeneous mice. Nature 2009; 460:392-5.

92. Zhang C, Li S, Yang L, Huang P, Li W, Wang S, et al. Structural modulation of gut microbiota in life-long calorie-restricted mice. Nat Commun 2013; 4:2163.

93. Walford RL, Harris SB, Gunion MW. The calorically restricted low-fat nutrient-dense diet in Biosphere 2 significantly lowers blood glucose, total leukocyte count, cholesterol, and blood pressure in humans. Proc Natl Acad Sci U S A 1992; 89:11533-7.

94. Fontana L, Meyer TE, Klein S, Holloszy JO. Long-term calorie restriction is highly effective in reducing the risk for atherosclerosis in humans. Proc Natl Acad Sci U S A 2004; 101:6659-63.

95. Fontana L, Villareal DT, Weiss EP, Racette SB, Steger-May K, Klein S, et al. Calorie restriction or exercise: effects on coronary heart disease risk factors. A randomized, controlled trial. Am J Physiol Endocrinol Metab 2007; 293:E197-202.

96. Heilbronn LK, de Jonge L, Frisard MI, DeLany JP, Larson-Meyer DE, Rood J, et al. Effect of 6-month calorie restriction on biomarkers of longevity, metabolic adaptation, and oxidative stress in overweight individuals: a randomized controlled trial. JAMA 2006; 295:1539-48.

97. Heidinger BJ, Blount JD, Boner W, Griffiths K, Metcalfe NB, Monaghan P. Telomere length in early life predicts lifespan. Proc Natl Acad Sci U S A 2012; 109:1743-8.

98. Menon R, Yu J, Basanta-Henry P, Brou L, Berga SL, Fortunato SJ, et al. Short fetal leukocyte telomere length and preterm prelabor rupture of the membranes. PLoS One 2012; 7:e31136.

99. Entringer S, Epel ES, Kumsta R, Lin J, Hellhammer DH, Blackburn EH, et al. Stress exposure in intrauterine life is associated with shorter telomere length in young adulthood. Proc Natl Acad Sci U S A 2011; 108:E513-8.

100. Theall KP, Brett ZH, Shirtcliff EA, Dunn EC, Drury SS. Neighborhood disorder and telomeres: connecting children's exposure to community level stress and cellular response. Soc Sci Med 2013; 85:50-8.

101. Eisenberg DT, Hayes MG, Kuzawa CW. Delayed paternal age of reproduction in humans is associated with longer telomeres across two generations of descendants. Proc Natl Acad Sci U S A 2012; 109:10251-6.

102. Aviv A. Genetics of leukocyte telomere length and its role in atherosclerosis. Mutat Res 2012; 730:68-74.

103. Prescott J, Du M, Wong JY, Han J, De Vivo I. Paternal age at birth is associated with offspring leukocyte telomere length in the nurses' health study. Hum Reprod 2012; 27:3622-31.

104. Hallows SE, Regnault TR, Betts DH. The long and short of it. J Pregnancy 2012; 2012:638476.

105. Kudo T, Izutsu T, Sato T. Telomerase activity and apoptosis as indicators of ageing in placenta with and without intrauterine growth retardation. Placenta 2000; 21:493-500.

106. Biron-Shental T, Sukenik-Halevy R, Sharon Y, Goldberg-Bittman L, Kidron D, Fejgin MD, et al. Short telomeres may play a role in placental dysfunction in preeclampsia and intrauterine growth restriction. Am J Obstet Gynecol 2010; 202:381 e1-7.

107. Davy P, Nagata M, Bullard P, Fogelson NS, Allsopp R. Fetal growth restriction is associated with accelerated telomere shortening and increased expression of cell senescence markers in the placenta. Placenta 2009; 30:539-42.

108. Hracsko Z, Orvos H, Novak Z, Pal A, Varga IS. Evaluation of oxidative stress markers in neonates with intra-uterine growth retardation. Redox Rep 2008; 13:11-6.

109. Chen RJ, Chu CT, Huang SC, Chow SN, Hsieh CY. Telomerase activity in gestational trophoblastic disease and placental tissue from early and late human pregnancies. Hum Reprod 2002; 17:463-8.

110. Kim SY, Lee SP, Lee JS, Yoon SJ, Jun G, Hwang YJ. Telomerase and apoptosis in the placental trophoblasts of growth discordant twins. Yonsei Med J 2006; 47:698-705.

111. Luyckx VA, Compston CA, Simmen T, Mueller TF. Accelerated senescence in kidneys of low-birth-weight rats after catch-up growth. Am J Physiol Renal Physiol 2009; 297:F1697-705.

112. Tarry-Adkins JL, Chen JH, Smith NS, Jones RH, Cherif H, Ozanne SE. Poor maternal nutrition followed by accelerated postnatal growth leads to telomere shortening and increased markers of cell senescence in rat islets. FASEB J 2009; 23:1521-8.

113. Jennings BJ, Ozanne SE, Dorling MW, Hales CN. Early growth determines longevity in male rats and may be related to telomere shortening in the kidney. FEBS Lett 1999; 448:4-8.

114. Menon R, Fortunato SJ, Milne GL, Brou L, Carnevale C, Sanchez SC, et al. Amniotic fluid eicosanoids in preterm and term births: effects of risk factors for spontaneous preterm labor. Obstet Gynecol 2011; 118:121-34.

115. Aycicek A, Erel O, Kocyigit A. Increased oxidative stress in infants exposed to passive smoking. Eur J Pediatr 2005.

116. Martin RM, Davey Smith G, Mangtani P, Tilling K, Frankel S, Gunnell D. Breastfeeding and cardiovascular mortality: The Boyd Orr cohort and a systematic review with meta-analysis. European Heart Journal 2004; 25:778–86.

117. Martin RM, Gunnell D, Smith GD. Breastfeeding in infancy and blood pressure in later life: systematic review and meta-analysis. Am J Epidemiol 2005; 161:15-26.

118. Rich-Edwards JW, Stampfer MJ, Manson JE, Rosner B, Hu FB, Michels KB, et al. Breastfeeding during infancy and the risk of cardiovascular disease in adulthood Epidemiology 2004; 15:550–6.

119. Anderson JW, Johnstone BM, Remley DT. Breast-feeding and cognitive development: A meta-analysis. American journal of clinical nutrition 1999; 70:525–35.

120. Der G, Batty GD, Deary IJ. Effect of breast feeding on intelligence in children: Prospective study, sibling pairs analysis, and meta-analysis. BMJ. 2006; 333:945.

121. Kramer MS, Aboud F, Mironova E, Vanilovich I, Platt RW, Matush L, et al. Breastfeeding and child cognitive development: New evidence from a large randomized trial. Archives of General Psychiatry 2008; 65:578-84.

122. Makrides M, Smithers LG, Gibson RA. Role of long-chain polyunsaturated fatty acids in neurodevelopment and growth. Nestle Nutrition workshop series. Paediatric programme 2010; 65:123-33; discussion 33-6.

Chapter 12: A new world of opportunity

1. Marmot M. Social determinants of health inequalities. Lancet 2005; 365:1099-104.

2. Wilkinson R, Marmot M. Social Determinants of Health: The Solid Facts (second edition). World Health Organization. 2003.

3. Berkman LF, Melchior M. The shape of things to come: How social policy impacts social integration and family structure to produce population health. In: Siegrist J, Marmot M. Social inequalities in health: New Evidence and Policy Implications. Oxford: Oxford University Press; 2006. pp55-72.

4. ibid.

5. Rose G. Strategy of preventive medicine. Oxford: Oxford University Press; 1992.

6. Daniell MH. World of Risk: A New Approach to Global Strategy and Leadership. Singapore: World Scientific Publishing Co.; 2004.

7. Achieving Aboriginal and Torres Strait Islander health equality within a generation - A human rights based approach. Social Justice Report, Australian Human Rights Commission (Chapter 2). 2005. [Cited October 2014] <https://www.humanrights.gov. au/publications/social-justice-report-2005-chapter-2-achieving- aboriginal-and-torres-strait-islander>.

8. Marmot M. Social determinants of health inequalities. Lancet 2005; 365:1099-104.

9. Progress and priorities report. The Close the Gap Campaign Steering Committee for Indigenous Health Equality, February 2014. [Cited October 2014] <https://www.humanrights.gov.au/sites/default/files/document/publication/ctg-progress-and-priorities-report.pdf>

10. Life Tables for Aboriginal and Torres Strait Islander Australians, 2010-2012, (cat. no 3302.0.55.003). Australian Bureau of Statistics. 2013.

11. Closing the Gap. Prime Minister's Report. Australian Government. 2014. [Cited October 2014] <http://www.dpmc.gov.au/publications/docs/closing_the_gap_2014.pdf>.

12. A statistical overview of Aboriginal and Torres Strait Islander peoples in Australia: Social Justice Report 2008, Australian Human Rights Commission. [Cited October 2014] <https://www.humanrights.gov.au/publications/statistical-overview-aboriginal-and-torres-strait-islander-peoples-australia-social>.

13. Marmot M. Social determinants of health inequalities. Lancet 2005; 365:1099-104.

14. Africa to benefit from Brazil-FAO school meals experience. Food and Agriculture Organization of the United Nations., 2013. [Cited October 2014] <http://www.fao.org/countryprofiles/news-article/en/c/202619/>.

15. Elliott T, Trevena H, Sacks G, Dunford E, Martin J, Webster J, et al. A systematic interim assessment of the Australian Government's Food and Health Dialogue. Med J Aust 2014; 200:92-5.

16. The Constitution of the Iroquois Nations. The Great Binding Law. Prepared by Gerald Murphy. National Public Telecomputing Network (NPTN).

Postscript

1. Slow down and enjoy the ride. Office of Road Safety. Government of Western Australia., 2013.

Appendix

1. Davies P, Funder J, Palmer DJ, Sinn J, Vickers M, Wall C. Early life nutrition. The opportunity to influence long-term health. 2014 [cited October 2014].< http://www.earlylifenutrition.org/pdf/EarlyLifeNutrition_FINAL.pdf>.

2. Royal Australian and New Zealand College of Obstetricians and Gynaecologists. Pre-pregnancy Counselling. College Statement C-Obs 3 (a); Current: November 2012.

3. Royal Australian and New Zealand College of Obstetricians and Gynaecologists. Management of Obesity in Pregnancy. College Statement C-Obs 49. Current: September 2013a.

4. Bammann K, Peplies J, De Henauw S, Hunsberger M, Molnar D, Moreno LA, et al. Early life course risk factors for childhood obesity: the IDEFICS case-control study. PLoS One 2014; 9:e86914.

5. Gluckman PD, Hanson MA, Cooper C, Thornburg KL. Effect of in utero and early-life conditions on adult health and disease. N Engl J Med 2008; 359:61-73.

6. National Institute of Health and Clinical Excellence. Weight management before, during and after pregnancy. NICE public health guidance no. 27. July 2010. [Cited 2014] <guidance.nice.org.uk/ph27>.

7. Painter RC, de Rooij SR, Bossuyt PM, Simmers TA, Osmond C, Barker DJ, et al. Early onset of coronary artery disease after prenatal exposure to the Dutch famine. Am J Clin Nutr 2006; 84:322-7; quiz 466-7.

8. Eriksson JG. Epidemiology, genes and the environment: lessons learned from the Helsinki Birth Cohort Study. J Intern Med 2007; 261:418-25.

9. National Institute of Health and Clinical Excellence. Weight management. [Cited 2014] <guidance.nice.org.uk/ph27>.

10. Rasmussen KM, Yaktine AL. Committee to Reexamine IOM Pregnancy Weight Guidelines. Weight gain during pregnancy: reexamining the guidelines. Washington: The National Academies Press. 2009.

11. Royal Australian and New Zealand College of Obstetricians and Gynaecologists. Pre-pregnancy Counselling. College Statement C-Obs 3 (a); Current: November 2012.

12. Royal Australian and New Zealand College of Obstetricians and Gynaecologists. Women and smoking. College Statement C-Obs 53. Current: November 2011.

13. Royal Australian and New Zealand College of Obstetricians and Gynaecologists. Pre-pregnancy Counselling. College Statement C-Obs 3 (a); Current: November 2012.

14. National Institute of Health and Clinical Excellence. Diabetes in pregnancy: Management of diabetes and its complications from pre-conception to the postnatal period. NICE clinical guideline 63. March 2008; [cited October 2014] <guidance.nice.org.uk/cg63>.

15. Royal Australian and New Zealand College of Obstetricians and Gynaecologists. Vitamin and Mineral Supplementation and Pregnancy. College Statement C-Obs 25; Current March 2013.

16. ibid.
17. Ministry of Health New Zealand. Food and Nutrition Guidelines for Healthy Pregnant and Breastfeeding Women: A background paper. Wellington: Ministry of Health. 2006. [cited October 2014] <http://www.health.govt.nz/publication/food-and-nutrition-guidelines-healthy-pregnant-and-breastfeeding-women-background-paper>.
18. National Health and Medical Research Council (2012) Infant Feeding Guidelines. Canberra: National Health and Medical Research Council. [cited October 2014] < http://www.nhmrc.gov.au/_files_nhmrc/publications/attachments/n56_infant_feeding_guidelines.pdf>.
19. Ministry of Health New Zealand. Food and Nutrition Guidelines 2006. [cited October 2014] <http://www.health.govt.nz/publication/food-and-nutrition-guidelines-healthy-pregnant-and-breastfeeding-women-background-paper>.
20. Food Standards Australia & New Zealand. Advice on fish consumption: Mercury in fish. 2011. [cited October 2014 <http://www.foodstandards.gov.au/consumer/chemicals/mercury/Pages/default.aspx>.
21. Royal Australian and New Zealand College of Obstetricians and Gynaecologists. Vitamin and Mineral Supplementation and Pregnancy. College Statement C-Obs 25; Current March 2013.
22. Ministry of Health New Zealand. Food and Nutrition Guidelines 2006. [cited October 2014] <http://www.health.govt.nz/publication/food-and-nutrition-guidelines-healthy-pregnant-and-breastfeeding-women-background-paper>.
23. ibid.
24. Stewart CP, Christian P, Schulze KJ, Arguello M, LeClerq SC, Khatry SK, et al. Low maternal vitamin B-12 status is associated with offspring insulin resistance regardless of antenatal micronutrient supplementation in rural Nepal. J Nutr 2011; 141:1912-7.
25. Royal Australian and New Zealand College of Obstetricians and Gynaecologists. Vitamin and Mineral Supplementation and Pregnancy. College Statement C-Obs 25; Current March 2013.
26. Ministry of Health New Zealand. Food and Nutrition Guidelines 2006. [cited October 2014] <http://www.health.govt.nz/publication/food-and-nutrition-guidelines-healthy-pregnant-and-breastfeeding-women-background-paper>.
27. Brunvand L, Quigstad E, Urdal P, Haug E. Vitamin D deficiency and fetal growth. Early Hum Dev 1996; 45:27-33.
28. Zhu K, Whitehouse AJ, Hart PH, Kusel M, Mountain J, Lye S, et al. Maternal vitamin D status during pregnancy and bone mass in

offspring at 20 years of age: a prospective cohort study. J Bone Miner Res 2014; 29:1088-95.

29. Royal Australian and New Zealand College of Obstetricians and Gynaecologists. Vitamin and Mineral Supplementation and Pregnancy. College Statement C-Obs 25; Current March 2013.

30. Ministry of Health New Zealand. Food and Nutrition Guidelines 2006. [cited October 2014] <http://www.health.govt.nz/publication/food-and-nutrition-guidelines-healthy-pregnant-and-breastfeeding-women-background-paper>.

31. National Health and Medical Research Council (2013b) Australian Eating Guidelines. Recommended number of serves for adults; 2013. [cited Octber 2014] <http://www.eatforhealth.gov.au/food-essentials/how-much-do-we-need-each-day/recommended-number-serves-adults>.

32. Australasian Society of Clinical Immunology and Allergy. Infant feeding advice, 2010. [cited October 2014] <http://www.allergy.org.au/content/view/350/287>.

33. Nauta AJ, Ben Amor K, Knol J, Garssen J, van der Beek EM. Relevance of pre- and postnatal nutrition to development and interplay between the microbiota and metabolic and immune systems. Am J Clin Nutr 2013; 98:586S-93S.

34. Nylund L, Satokari R, Nikkila J, Rajilic-Stojanovic M, Kalliomaki M, Isolauri E, et al. Microarray analysis reveals marked intestinal microbiota aberrancy in infants having eczema compared to healthy children in at-risk for atopic disease. BMC Microbiol 2013; 13:12.

35. Amarasekera M, Prescott SL, Palmer DJ. Nutrition in early life, immune-programming and allergies: the role of epigenetics. Asian Pac J Allergy Immunol 2013; 31:175-82.

36. Laitinen K, Poussa T, Isolauri E. Probiotics and dietary counselling contribute to glucose regulation during and after pregnancy: a randomised controlled trial. Br J Nutr 2009; 101:1679-87.

37. Osborn DA, Sinn JK. Prebiotics in infants for prevention of allergy. Cochrane Database Syst Rev 2013; 3:CD006474.

38. National Health and Medical Research Council (2012) Infant Feeding Guidelines. Canberra. [cited October 2014] < http://www.nhmrc.gov.au/_files_nhmrc/publications/attachments/n56_infant_feeding_guidelines.pdf>.

39. ibid.

40. ibid.

41. ibid.

42. Ministry of Health New Zealand. Food and Nutrition Guidelines 2006. [cited October 2014] <http://www.health.govt.nz/publication/food-and-nutrition-guidelines-healthy-pregnant-and-breastfeeding-women-background-paper>.

43. National Health and Medical Research Council (2012) Infant Feeding Guidelines. Canberra: National Health and Medical Research Council. [cited October 2014] < http://www.nhmrc.gov.au/_files_nhmrc/publications/attachments/n56_infant_feeding_guidelines.pdf>.

44. National Health and Medical Research Council (2013b) Australian Eating Guidelines. Recommended number of serves for adults; 2013. [cited Octber 2014] <http://www.eatforhealth.gov.au/food-essentials/how-much-do-we-need-each-day/recommended-number-serves-adults>.

45. Australasian Society of Clinical Immunology and Allergy. Infant feeding advice, 2010. [cited October 2014] <http://www.allergy.org.au/content/view/350/287>.

46. Du Toit G, Katz Y, Sasieni P, Mesher D, Maleki SJ, Fisher HR, et al. Early consumption of peanuts in infancy is associated with a low prevalence of peanut allergy. J Allergy Clin Immunol 2008; 122:984-91.

47. Prescott SL, Smith P, Tang MLK, Palmer DJ, Sinn J, Huntley SJ, et al. The importance of early complementary feeding in the development of oral tolerance: concerns and controversies. Pediatr Allergy Immunol 2008; 19(5) 375-80.

48. Koplin JJ, Osborne NJ, Wake M, Martin PE, Gurrin LC, Robinson MN, et al. Can early introduction of egg prevent egg allergy in infants? A population-based study. J Allergy Clin Immunol 2010; 126:807-13.

49. Australasian Society of Clinical Immunology and Allergy. Infant feeding advice, 2010. Available at[cited October 2014] <http://www.allergy.org.au/content/view/350/287>.

50. National Health and Medical Research Council (2012) Infant Feeding Guidelines. Canberra: National Health and Medical Research Council. [cited October 2014] < http://www.nhmrc.gov.au/_files_nhmrc/publications/attachments/n56_infant_feeding_guidelines.pdf>.

51. ibid.

52. Davies P, Funder J, Palmer DJ, Sinn J, Vickers M, Wall C. Early life nutrition. The opportunity to influence long-term health. 2014 [cited October 2014].< http://www.earlylifenutrition.org/pdf/EarlyLifeNutrition_FINAL.pdf>.

Index

rheumatoid arthritis (RA) 154,
175–9
Rousseau, Jean-Jacques 40
Royal Perth Hospital 6
rubella 34–5

schizophrenia
brain autopsies to discover 123
Dutch Hunger Winter and 67,
146
epigenetics and 135, 137
immune abnormalities in 140
inflammation and 140–1
inflammation in pregnancy and
32
medical imaging to discover 123
neuronal pruning and 139
prevalence 125
'two-hit hypothesis' 141–2
scoliosis 168–9
Semmelweis, Ignaz 42
Shanghai Declaration (UN) 17–18
short-chain fatty acids (SCFA) 56,
226–7
silymarin 199
Simmer, Karen (Professor) 9
single nucleotide polymorphisms
(SNPs) 246, 249
sitting time 52, 55, 174
skin cancer 201–2
sleep disturbance
ageing effects 244
cardiovascular disease and 113–14
inflammation and 114
obesity and 53
obstructive sleep apnoea 114
slipped capital femoral epiphysis
(SCFE) 170, 180
smoking
cancer and 196–7
inflammation from 109
paternal 287

in pregnancy 52, 103, 117, 176,
196–7, 216, 222, 260, 287
rheumatoid arthritis 179
risks from early exposure 20–1
social determinants of health 14, 24,
27, 71, 118–9, 133, 171, 263–6
*Social Determinants of Health: The
Solid Facts* (WHO) 263
soft drinks 88, 90–1, 163, 167
solid food, introducing 69–71,
228–30, 295–6
spina bifida 34–5, 126–7, 146, 288
stannous chloride (SnCl2) 160
statistics *see* global statistics
steroid medications 162
stomach cancer 199–200
Straker, Leon (Professor) 173
stress
ageing and 264cancer and 188
exercise to reduce 55
early nurturing and 131, 133, 138
hypertension and 101–02
immune system and 139–41
in pregnancy 30, 33, 104, 159,
223, 258–60
obesity and 53
reduction 150–51, 280–3
response 55, 130–31, 133–6, 138,
141, 147–8, 159, 216
schizophrenia and 142
telomere shortening and 244,
258–60
yoga and 102
substance-use disorders 125
sugar 12, 52, 79, 81, 88, 90–3, 102,
166–7, 269
suicide 136, 216
sulphoraphane 199
symbionts 219, 225–6
synaptic pruning 129–31, 139
systemic 'lupus' erythematosus
(SLE) 175–9

www.ingramcontent.com/pod-product-compliance
Lightning Source LLC
Chambersburg PA
CBHW020332270326
41926CB00007B/151